Revision for SCIENCE *Key Stage 3*

• with answers •

REVISED NATIONAL CURRICULUM EDITION

JOE BOYD WALTER WHITELAW

JOHN MURRAY

Revision Guides:

Revision for Science Key Stage 3 with Answers	0 7195 7249 5
Revision for Science Key Stage 4 with Answers Revised National Curriculum Edition	0 7195 7422 6
Revision for Maths Levels 3–8 Intermediate & GCSE with Answers	0 7195 7083 2
Revision for English Key Stage 3	0 7195 7025 5
Revision for History GCSE Modern World History	0 7195 7229 0
Revision for French GCSE with Answers and Cassette	0 7195 7306 8
Revision for German GCSE with Answers and Cassette	0 7195 7309 2
Revision for Spanish GCSE with Answers and Cassette	0 7195 7394 7

First published in 1994
by John Murray (Publishers) Ltd
50 Albemarle Street
London W1X 4BD

Reprinted in 1995

Second edition 1996
Reprinted 1996, 1998, 1999

Layouts by Steve Rowling
Illustrations by Technical Art Services
Typeset in 11/12½pt New Century Schoolbook
Printed in Great Britain by St Edmundsbury Press Ltd, Bury St Edmunds, Suffolk

A CIP catalogue entry for this book is available from the British Library

ISBN 0-7195-7249-5

Contents

Introduction

During the last few years, you have gained a great deal of important knowledge and understanding from the National Curriculum for Science. You have probably forgotten some of the work. This work must be revised and relearned for your examinations. Unfortunately, revision is hard work and it is easy to avoid. Some of the most common ways are shown in the cartoons below.

First, get organised!

Plan

It is always a good idea to plan your revision. The plan below is an example of a revision timetable. Each box represents 2 or 3 hours of work. There is one box for a school day, and two for weekend and holiday days. Make up something like this for your own revision, and stick it on your wall.

Study/Revision planner

1. Each box represents 2 or 3 hours of work.
2. Plan all your revision before you begin it.
3. Try to keep to your plan, but don't be afraid to change it.

Mon	Tues	Wed	Thurs	Fri	Sat am	Sat pm	Sun am	Sun pm
/	/	/	/	20 March	Sc 2 level 3	Sc 2 level 4	Day off	Day off
Sc 3 level 3	Sc 3 level 4	Sc3 level 5	etc	27 March				
				3 April				
				10 April				

Be an active learner

Try to be an active learner. Rather than just read through notes, make sure that you write or draw or underline as you read through them. Make summaries and answer questions. Use spider diagrams and revision summaries (examples of these are given below).

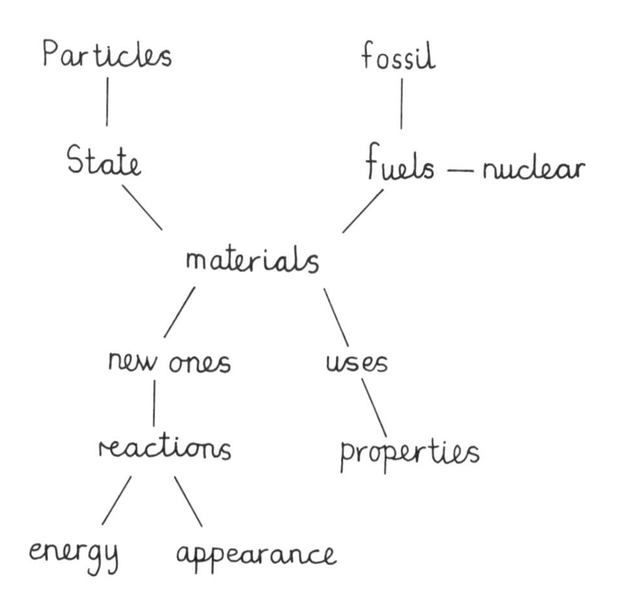

Example of a spider diagram

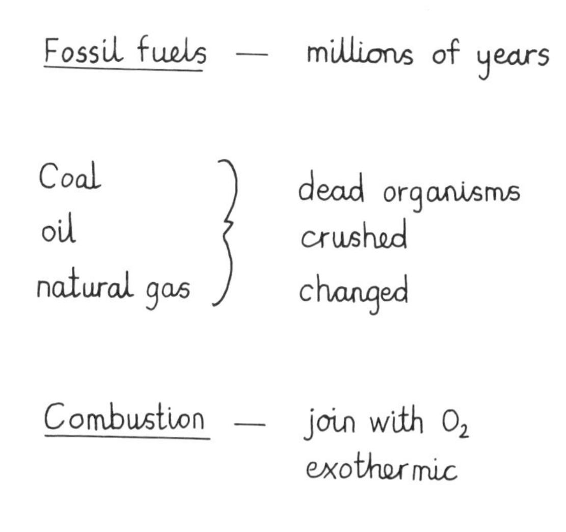

Example of a revision summary

Next, use this book!

Get your own copy

This book contains the main ideas of the whole Science course, Key Stage 3, Levels 3–7. Your teacher will suggest which level you should aim for. Learn the levels in order, from 3 upwards, until you reach your own target. If you are aiming for Level 5, begin with Level 3, then learn 4, then 5.

Write all over it

1. Each section starts by listing the main 'big' ideas. Read these and make sure you understand the sense of the work. The paragraph letters link with each big idea.
2. Read each paragraph. In early levels, write the key words from the paragraph in the margin. In later levels, underline the key phrases.
3. Continue with the other paragraphs. Complete any diagrams when asked.
4. At the end of each double page, read over your work. Tick the boxes when you know and understand the work.
5. Check your answers if a check sheet is available. (See your teacher about anything you still do not understand.)
6. Have a short five-minute break.
7. Start the next double page of work.

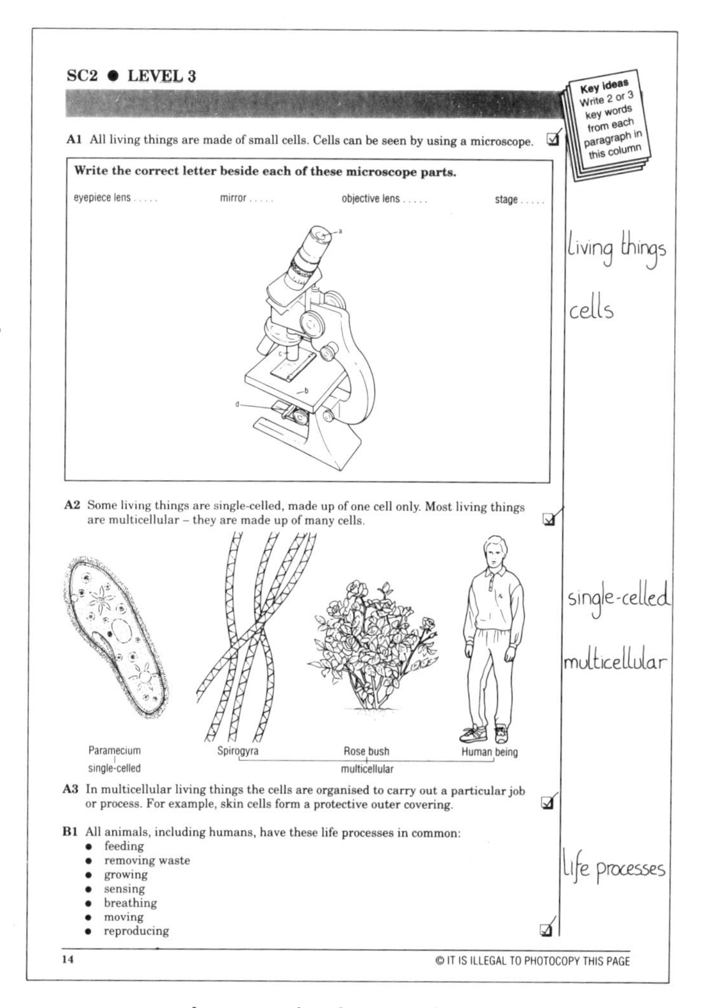

SC2 ● LEVEL 3

Key ideas
Write 2 or 3 key words from each paragraph in this column

A1 All living things are made of small cells. Cells can be seen by using a microscope.

Write the correct letter beside each of these microscope parts.

eyepiece lens ..... mirror ..... objective lens ..... stage .....

living things
cells

A2 Some living things are single-celled, made up of one cell only. Most living things are multicellular – they are made up of many cells.

single-celled
multicellular

A3 In multicellular living things the cells are organised to carry out a particular job or process. For example, skin cells form a protective outer covering.

B1 All animals, including humans, have these life processes in common:

- feeding
- removing waste
- growing
- sensing
- breathing
- moving
- reproducing

life processes

14 © IT IS ILLEGAL TO PHOTOCOPY THIS PAGE

An example of a completed page

Finally, record your revision

Complete this record sheet after each revision session. It will help you to plan your work.

Attainment Target		Session 1 Date completed	Session 2 Date completed	Session 3 Date completed
1 Experimental and Investigative Science	Level 3			
	Level 4			
	Level 5			
	Level 6			
	Level 7			
2 Life Processes and Living Things	Level 3			
	Level 4			
	Level 5			
	Level 6			
	Level 7			
3 Materials and their Properties	Level 3			
	Level 4			
	Level 5			
	Level 6			
	Level 7			
4 Physical Processes	Level 3			
	Level 4			
	Level 5			
	Level 6			
	Level 7			

ATTAINMENT TARGET 1

Experimental and Investigative Science

Big ideas

You should be able to plan and carry out investigations in which you:

A ask questions, predict and plan experiments
B observe, alter variables, measure changes and record evidence
C present and interpret results and draw conclusions
D evaluate evidence.

These big ideas need to be explained more fully. The words down the left-hand side are about all your investigations. The meanings of these words are given on the right-hand side.

A	ask	• use your knowledge and ideas to ask a sensible question which can be investigated
	predict	• work out what is likely to happen
	plan	• design an experiment which is a fair test
B	observe	• look carefully at what happens when you do your test • identify important variables: these are things which could affect the results
	alter	• change one variable deliberately • keep the other variables the same
	measure	• measure the effect of this change accurately • repeat the measurements if you need to check their accuracy
	record	• write down all your results
C	present	• use tables and graphs to show your results
	interpret	• think about the results • try to find a pattern in the results to explain them
	conclude	• decide if the results support your original prediction
D	evaluate	• judge how good the investigation was

Each level is illustrated by one investigation.

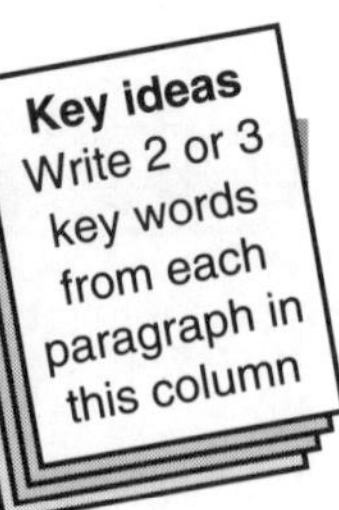

Ask: what helps cress grow?

A1 Ideas

Growth might depend on:
the amount of water
the amount of light
the amount of nitrogen
the temperature

Predictions

more water will increase growth
no light will stop growth
more nitrogen will increase growth
..too..hot..or..too..cool.....
..will..stop..growth.........

Investigation: Will cress seedlings grow with no light? ❑

B1 The following **variables** are all likely to affect growth:

- light
- amount of water
- type of seeds
- temperature
- amount of space (write in two other variables)

Alter one variable, e.g. *amount of light*. Keep all the others the same. Leave for a reasonable time. **Observe** the appearance, then **measure** the height of the seedling.

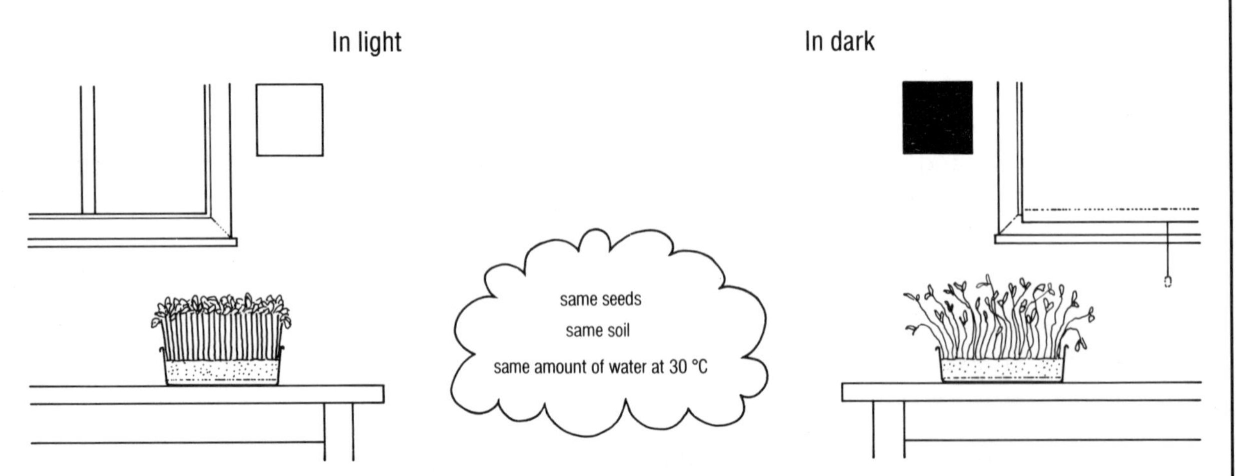

❑

C1 The results need to be **interpreted**. Although the height of the cress seedling is greater without the light, the plant does not look healthy; it is yellow and spindly. In general, our results suggest that cress seedlings grow more healthily in the light than in the dark.

C2 We can **conclude** that:
light is needed for cress to grow properly. ❑

Key ideas
Write 2 or 3 key words from each paragraph in this column

A1 Ask: how can you keep water hot in a container?

Ideas	**Predictions**
Heat kept in might depend on:	
the insulating material	wool better than wood, etc.
starting temperature of water	higher temperature loses heat faster
the thickness of insulation	increased thickness is best
....	....
....	....

Investigation: Will heat loss decrease as the thickness of the insulating material increases? ❑

B1 The following **variables** will all affect the amount of heat lost:

- material used
- thickness of material
-
-                          (write in two other variables)

Alter one variable, e.g. *thickness of layer.* Keep all the others the same. Set up a range of experiments. Leave for (say) 10 minutes. Then **measure** the temperature of the water.

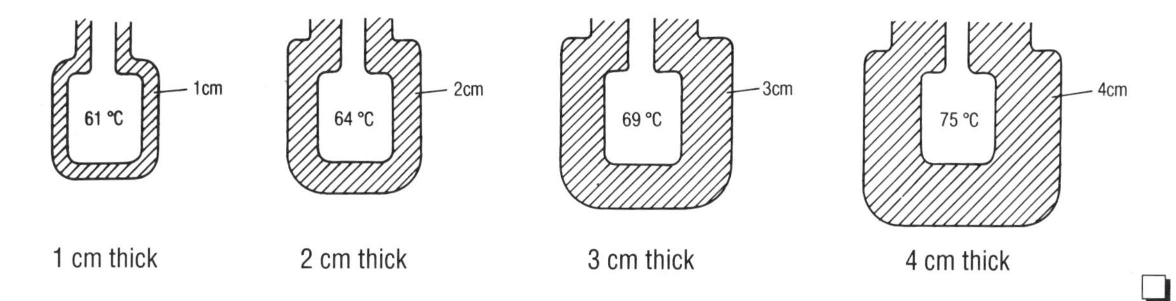

❑

C1 The results need to be **interpreted**. The temperature has dropped in all cases. However, it has decreased by different amounts. Sometimes a graph will show up the relationship between the variables very clearly.

Sketch the likely graph on these axes.

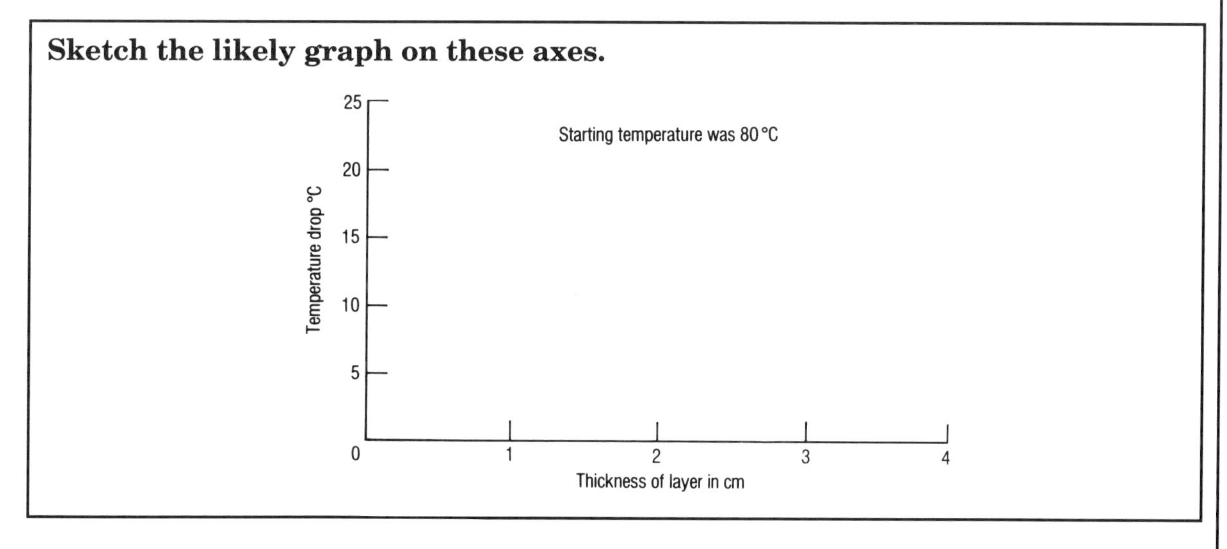

C2 From the graph we can **conclude** that:
the thicker the layer of insulating material, the less heat will be lost from the container. ❑

Key ideas
Write 2 or 3 key words from each paragraph in this column

Ask: what affects the rate at which bread dough rises?

A1 Your own scientific knowledge is useful when it comes to planning this investigation. From your studies you know that:

- a living plant called yeast causes dough to rise
- factors such as temperature and concentration of materials are important in changing the rate of a chemical reaction.

Ideas	**Predictions**
the rate might depend on:	
temperature	high temperature will increase rate but high temperature may kill yeast
concentration of materials	more sugar will increase rate
shape of container	wide opening will decrease rate
....	....
....	....

Investigation: What is the best temperature for the living yeast to make the dough rise quickly? ❑

B1 The following **variables** will all affect the rate at which the bread dough rises:

- materials used
- yeast used
- temperature
-
- (write in two other variables)

Alter one variable, e.g. *temperature.* Keep all the others the same. Set up a range of experiments and leave for 15 minutes. Choose a range of temperatures, say from room temperature to about 60 °C because you do not expect the yeast to survive outside this range. **Measure** the height of rise of the dough.

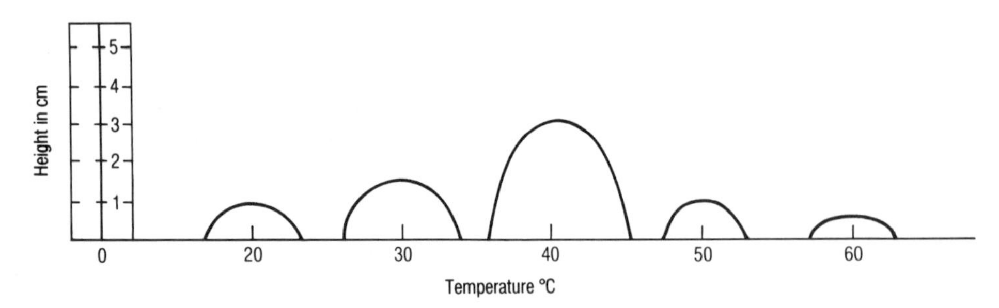

C1 The results need to be **interpreted**. The dough rose in all cases. However, it rose by different amounts. Sometimes a graph will show up the relationship between the variables very clearly.

C2 From the graph we can **conclude** that: **the yeast makes the dough rise highest at a temperature of about 40 °C.** ❑

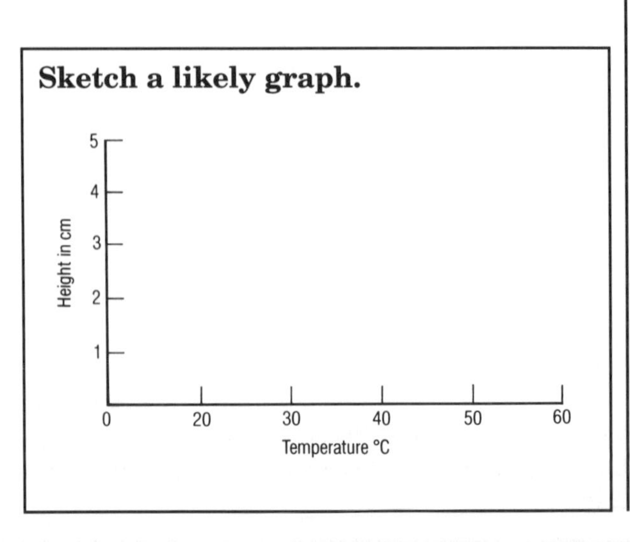

Underline or highlight key phrases.

Ask: what factors affect the rate of a chemical reaction?

A1 Your own scientific knowledge is useful when it comes to planning this investigation. From your studies, you can choose a suitable chemical reaction; the reaction between calcium carbonate and hydrochloric acid is a useful one because it produces a gas that is easy to collect and measure.

Ideas	**Predictions**
the rate might depend on:	
temperature	high temperature will increase rate
use of a catalyst	catalyst will increase rate
concentration of reactants	more acid will ....
particle size	crushed calcium carbonate will ....

Investigation: Does the rate increase as the concentration of acid increases?

B1 The following **variables** will all affect the rate of reaction:

- concentration of acid
- temperature
-
- (write in two other variables)

Alter one variable, e.g. *concentration.* Keep all the others the same. Set up a range of experiments. Choose a range of concentrations to give a safe and reasonable reaction. (Sometimes it is necessary to pre-test your experiments to work out the correct range for an investigation.) **Measure** the time taken to lose 1 g of gas.

0.5 M

1.0 M

1.5 M

2.0 M

C1 The results need to be **interpreted**. The weight loss was the same in all cases but the time taken to lose it was different. Sometimes a graph will show up the relationship between the variables very clearly.

Sketch the likely graph.

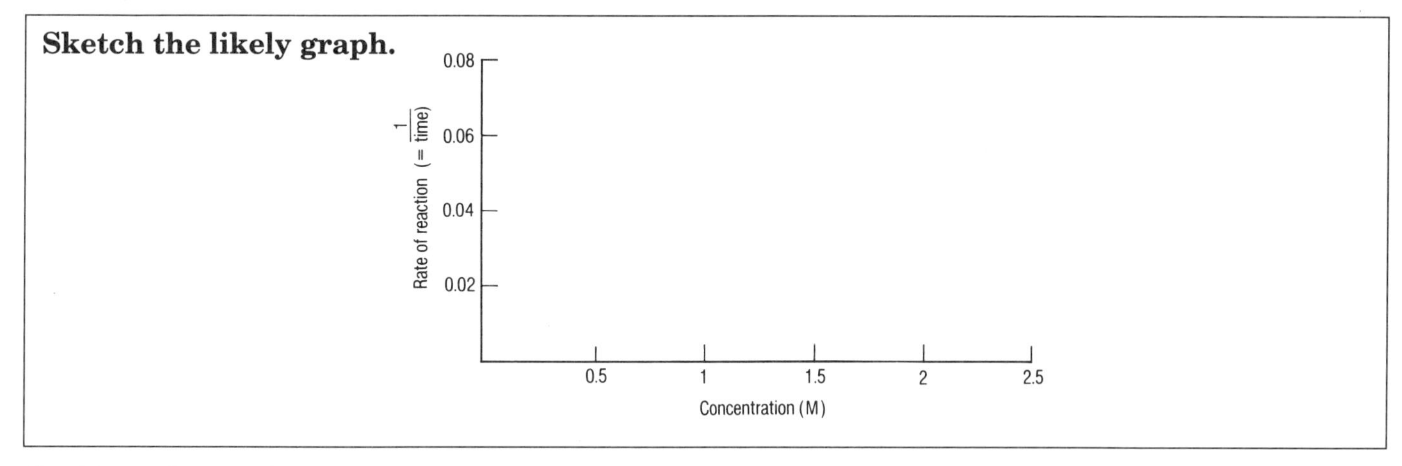

C2 From the graph we can **conclude** that:
the rate of reaction increases as the concentration of acid increases.
We can try to explain this by referring to a theory. Perhaps the rate has increased because there are more collisions between the reacting particles when the beaker is crowded with acid particles.

Underline or highlight key phrases.

Ask: why do plants of the same species in different places have leaves of different sizes?

A1 Sometimes an investigation has to deal with very complicated situations. It is likely in this example that a number of variables are contributing to the change in the size of leaves. In this real situation you cannot control all these variables. Instead, you have to measure as many variables as possible and then interpret your results.

Ideas
The size of leaf might depend on:
temperature, light level, soil depth, soil pH, moisture content

....

Prediction

The prediction is a general one: that there is a best value for each of the variables. At this optimum value, the plant will grow the biggest leaves

Investigation: Which variables influence the size of the plant's leaves? ❑

B1 The following **variables** may affect the size of leaf:
- soil depth, soil pH, temperature
-
- (write in other variables)

Keep the variable, type of plant, the same. **Measure** the value of all the variables. Also **measure** the size of the leaf. Do this in several different places. ❑

soil depth 5cm, pH6, light 7

Site A Open meadow

soil depth 6cm, pH7, light 2

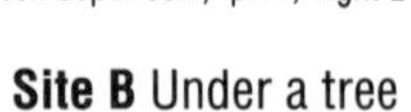

Site B Under a tree

soil depth 2cm, pH6, light 3

Site C On a shaded river bank

C1 The results need to be **interpreted**. It is likely that all the factors have an effect. However, the soil pH seems to have less of an effect than the light intensity. ❑

Examine the results and write down three conclusions from them.

D1 The results need to be **evaluated**. Have enough measurements been collected at each site? Were the measurements accurate? Could the investigation be improved?. ❑

ATTAINMENT TARGET 2

Life Processes and Living Things

Big ideas

A Plants and animals are made up of cells organised to carry out important life processes.

B Humans are organisms which carry out life processes.

C The life processes of green plants are important for our survival.

D Living things can be put into different groups. Some differences between individuals of the same type are inherited.

E Living things are adapted to survive in their surroundings but many things can affect the balance and numbers of living things in an ecosystem.

Key ideas
Write 2 or 3 key words from each paragraph in this column

A1 All living things are made of small cells. Cells can be seen by using a microscope. ❑

Write the correct letter beside each of these microscope parts.

eyepiece lens . . . . .   mirror . . . . .   objective lens . . . . .   stage . . . . .

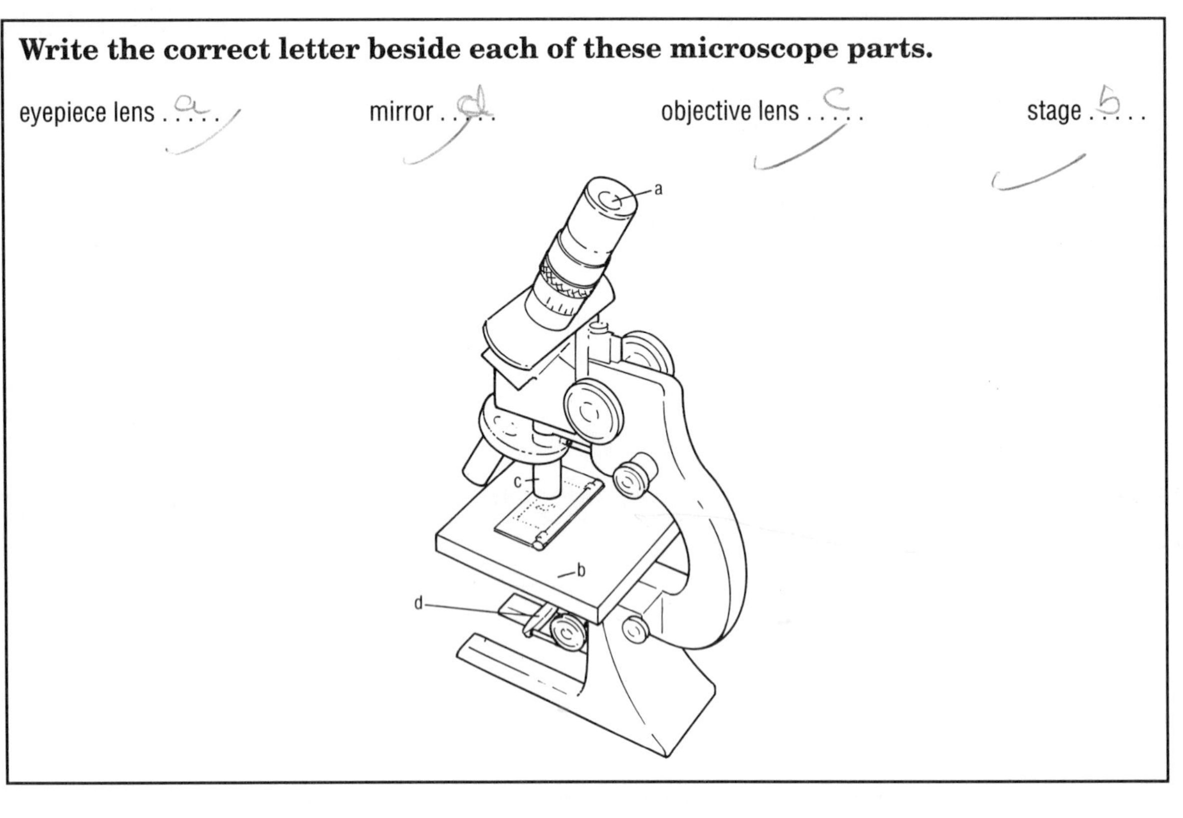

A2 Some living things are single-celled, made up of one cell only. Most living things are multicellular – they are made up of many cells. ❑

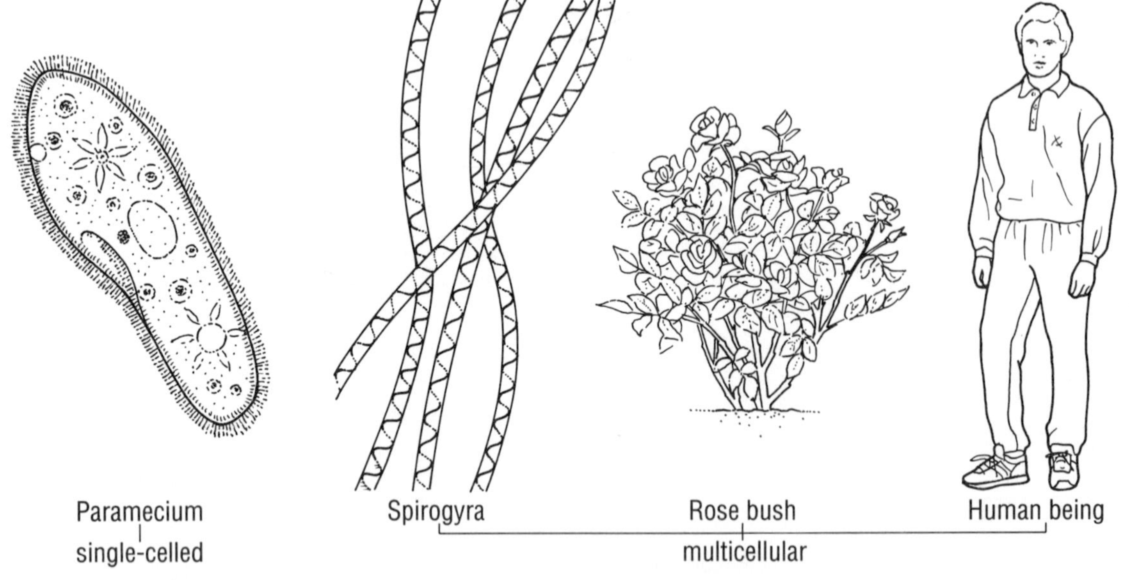

A3 In multicellular living things the cells are organised to carry out a particular job or process. For example, skin cells form a protective outer covering. ❑

B1 All animals, including humans, have these life processes in common:

- feeding
- removing waste
- growing
- sensing
- breathing
- moving
- reproducing

❑

Key ideas
Write 2 or 3 key words from each paragraph in this column

Write the name of the process in each space to link with the correct 'What it means' explanation.

Life process	What it means
reproducing	when a male and female produce new individuals
sensing	when a living thing is aware of changes around it and responds to these
feeding	when an animal takes in food for energy
growing	when an animal becomes bigger
breathing	when an animal takes in oxygen from the surrounding air
moving	when an animal goes from one place to another
removing waste	when an animal passes poisonous or unwanted material out of its body

B2 Non-living things may appear to be alive. For example, a falling rock moves and a crystal grows. However, a non-living (or a dead) thing does not show evidence of **all** the life processes ❑

C1 Green plants do not feed. They use light energy from the sun to stay alive and healthy. This life process is called **photosynthesis**. ❑

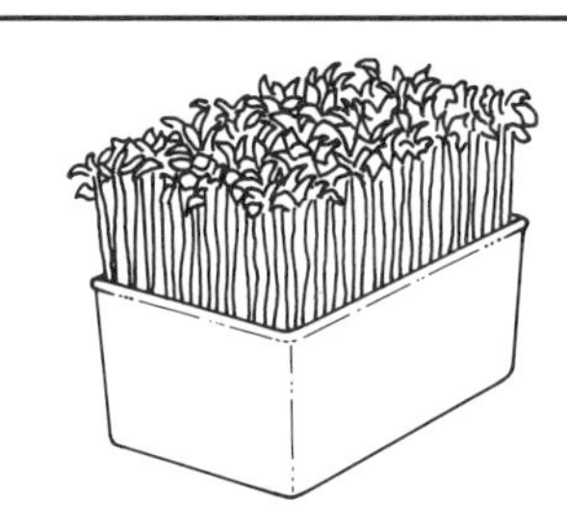

Cress grown in sunlight

Draw the result.

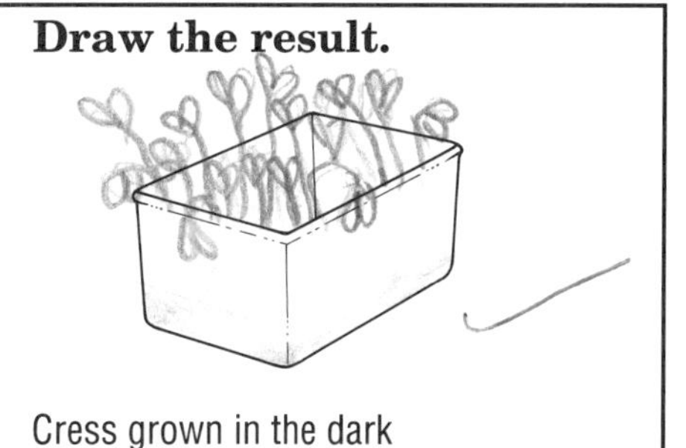

Cress grown in the dark

D1 **Variation** is the word used to describe the differences between living things. For example, an eagle, a swan and a sparrow show variation in beaks and feet. ❑

E1 The area around us is called the **environment**. ❑

E2 Animals are adapted (suited) to their environments. ❑

Complete the table. The first one is done for you.
Write down two other examples that you know about.

Animal	Environment	Example of adaptation
fish	water	fins for swimming
polar bear	freezing temperatures	thick fur
camel	dry deserts	humps for storing water
chameleon	trees	camoflauge for protection
duck	water	webbed feet for swimming

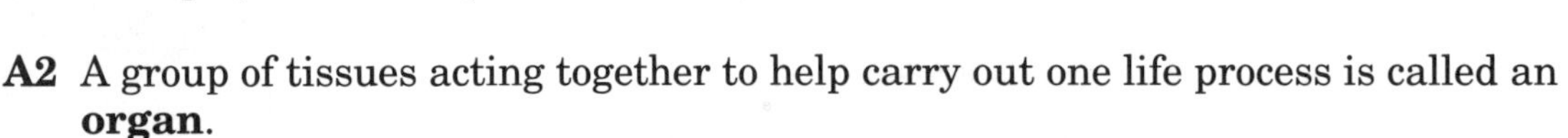

Key ideas
Write 2 or 3 key words from each paragraph in this column

A1 A group of cells working together to carry out the same job is called a **tissue**. For example, skin tissue, muscle tissue, bone tissue. ❑

A2 A group of tissues acting together to help carry out one life process is called an **organ**. ❑

Write the correct letter beside each human organ.

brain ..a.. heart ..c..

kidneys ..e.. stomach ..d..

lungs ..b..

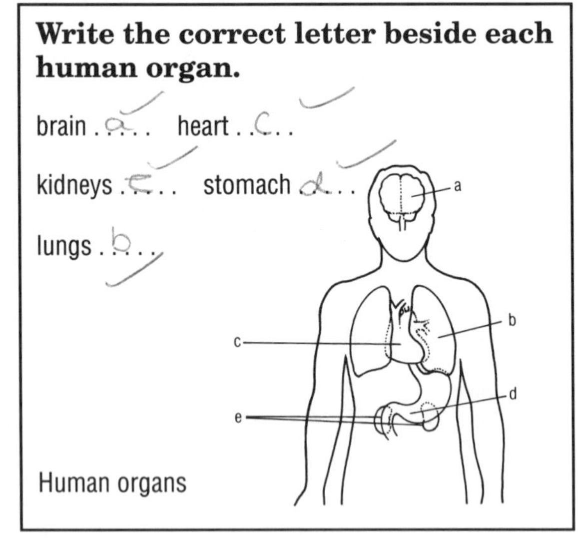

Human organs

Write the correct letter beside each plant organ.

flower ..d.. leaf ..c..

roots ..a.. stamen ..e..

ovary ..f..

stem ..b..

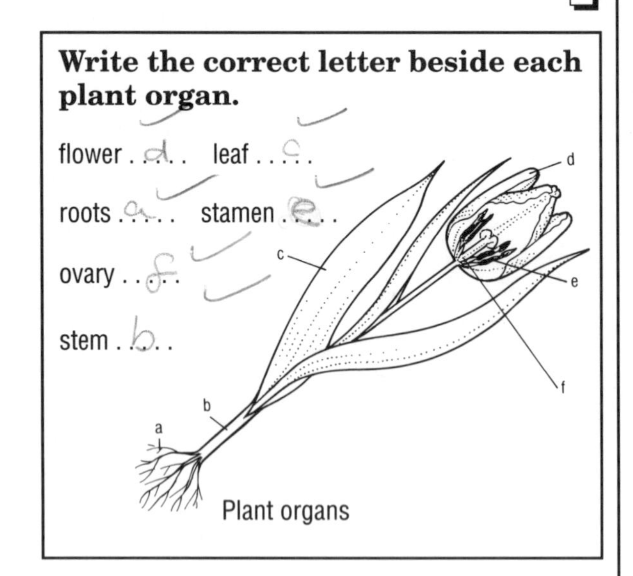

Plant organs

B1 Good health depends on eating a balanced diet, having healthy teeth and leading a healthy lifestyle by avoiding tobacco, alcohol and drugs. ❑

B2 Having a balanced diet means getting the correct amount of these components: protein, fat, carbohydrate, minerals, vitamins, fibre, water. ❑

Write down the three components of a balanced diet that can provide energy.

carbohydrates ....fat.......... ....protein..........

B3 Some foods are good sources of one or more components of a balanced diet. ❑

Complete the table giving two sources for each component – one from the list and one that you think of yourself.

Component	Sources	List
protein	eggs, chicken	tea
fat	fish liver oil, cheese	sugar
carbohydrate	sugar, potatoes	milk
minerals	milk	eggs
vitamins	fresh fruit, vegetables	fresh fruit
fibre	nuts, cereal	fish liver oil
water	tea, lettuce	nuts

C1 Flowering plants reproduce sexually. ❑

Put the following stages in the sexual reproduction of flowering plants in the correct order, labelled 1–4.

..1.. pollen lands on stigma of flower

..3.. male nucleus fertilises egg cell

..4.. ovule develops into seed

..2.. pollen tube grows down style to ovule

pollen
stigma
style
pollen tube
ovary
ovule
egg cell

Key ideas
Write 2 or 3 key words from each paragraph in this column

D1 A biological key can be used to identify and name organisms.

Write the name of each animal in the correct space.

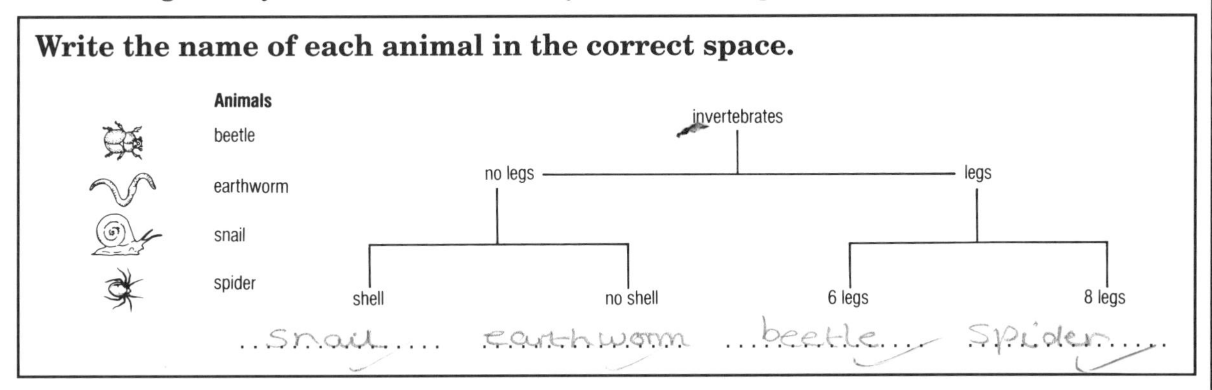

D2 The two main groups of animals are the **vertebrates** (animals with backbones) and the **invertebrates** (animals without backbones).

Circle the vertebrate animals and draw in their backbones.

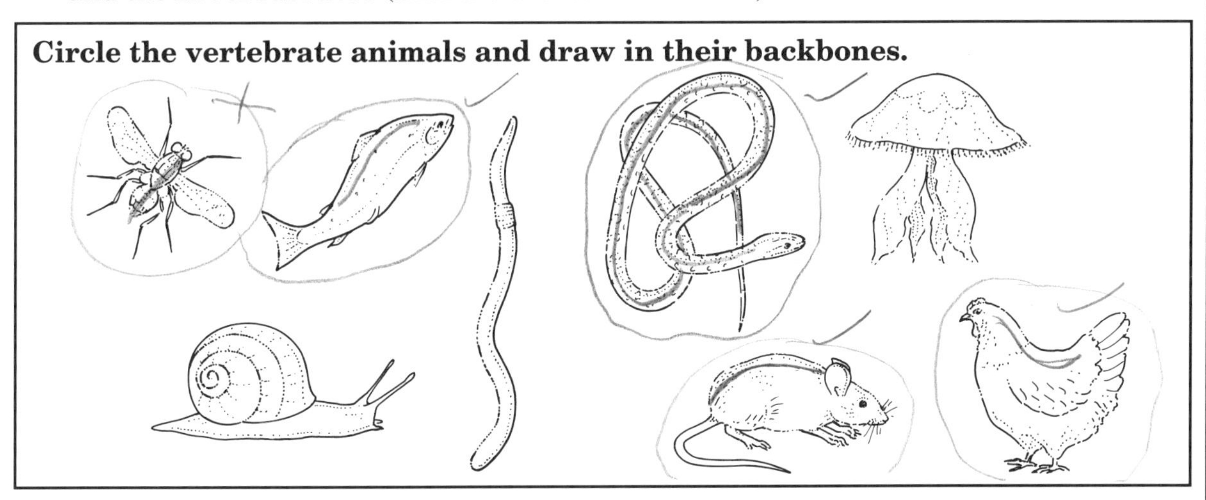

D3 Vertebrate animals can be classified into five sets with common features. These are: mammals, birds, reptiles, amphibians and fish.

Complete this table.

Name of vertebrate set	Body covering	Body temperature	Reproduction
amphibians	moist skin	cold blooded	egg laying
birds	feathers, scales	warm blooded	egg laying
fish	scales	cold blooded	egg laying
mammals	dry skin	warm blooded	live birth
reptiles	dry scales	cold blooded	egg laying

Key ideas
Write 2 or 3 key words from each paragraph in this column

D4 The two main groups of plants are flowering plants and non-flowering plants. ❑

grasses in flower
chestnut tree in flower
rose bush in flower
orchid in flower

Some flowering plants

D5 Non-flowering plants can be classified into smaller groups with common features:

- mosses are small plants with no true leaves or stems
- ferns are plants with leaves divided into tiny leaflets
- conifers are plants with seeds held in cones ❑

E1 In any environment, living things are in competition for scarce resources. ❑

Complete the sentence below.
Plants compete for space, light, water and minerals from the soil. ❑

Plants on the woodland floor must grow and flower before the bushes and trees above them produce leaves and cut out the light.

E2 Animals compete for space, food and water. When there are too many animals in one place many will die of starvation. ❑

E3 The environment, with all its plants and animals, is called an **ecosystem.** ❑

E4 A food chain shows the flow of energy through an ecosystem. ❑

E5 A food chain shows what eats what (the feeding relationships) in an ecosystem. The pattern in a food chain is always the same: ❑

energy from the sun → green plant → plant-eating animal → animal-eating animal → larger animal-eating animal

Tree
Caterpillar
Thrush
Hawk

A simple food chain

E6 Animals that are plant eaters are called herbivores. Animals that are meat eaters are called carnivores. An animal which kills and eats other animals is called a predator. The animal which is eaten is called the prey. ❑

Circle the herbivore and underline the carnivores in the food chain above.
Which two animals are predators? thrush and hawk
Which two animals are prey? caterpillar and thrush

Key ideas
Write 2 or 3 key words from each paragraph in this column

A1 Organs and tissues are arranged as organ systems in vertebrates. Most organ systems carry out one life process. ❑

B1 No single organ system controls growth. Growth takes place in many parts of a vertebrate's body. ❑

B2 The heart, blood vessels and blood make up the transport system. The transport system is involved in all life processes. ❑

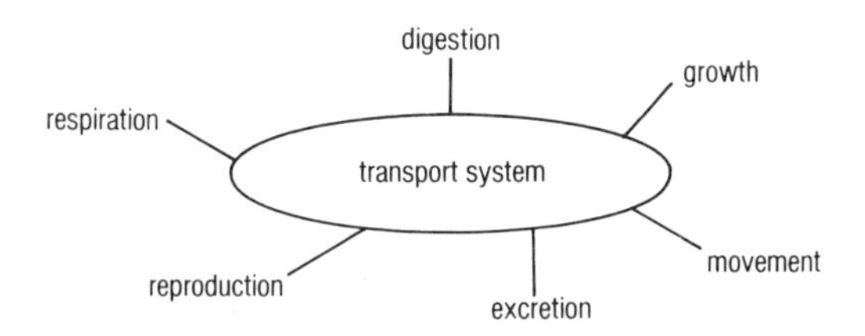

B3 Tiny blood vessels called capillaries deliver food and oxygen to every body cell. Food and oxygen pass out of the capillary and are exchanged for waste such as carbon dioxide. ❑

Label the blood capillary system in the diagram below.

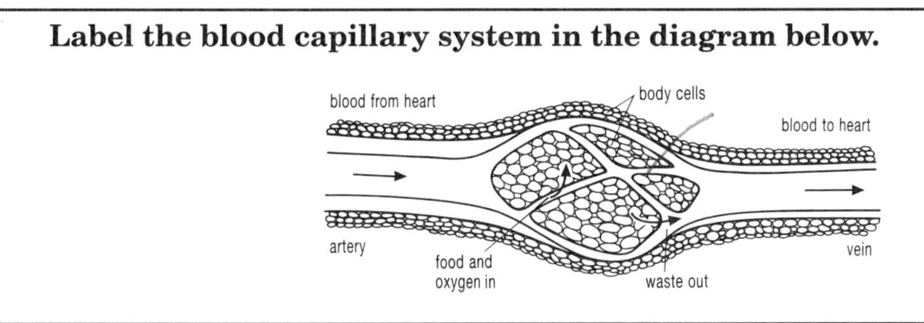

B4 Each organ system has a particular function. ❑

Write the name of the correct organ system beside each function. Choose from the following list: ~~respiratory~~, ~~excretory~~, ~~digestive~~, ~~skeletal~~, ~~reproductive~~, ~~transport~~, ~~nervous~~.

..transport.....	The heart pumps blood in blood vessels round the body. Blood carries food and oxygen to all cells of the body and removes waste. ✓
skeletal......	Bones act like levers. They move because the muscles attached to them contract (get smaller). ✓
digestive........	Large, insoluble food particles are broken down into small, soluble particles by enzymes. This takes place mostly in the small intestines. Digested food is passed to the blood supply. Undigested food moves along the large intestine and is egested at the anus. ✓
respiratory.....	Air passes down the windpipe to the lungs where oxygen is exchanged for carbon dioxide and water vapour. Movement of the diaphragm and ribs inflates and deflates the lungs. ✓
nervous.........	The brain collects information sent to it by certain nerves, and responds by sending messages to muscles. ✓
reproductive....	Sperm produced in the testes of the male are passed into the vagina of the female. The sperm swim through the womb (uterus) to the egg tube (fallopian tube). If an egg released by the ovary is present, it may be fertilised by one sperm. ✓
excretory.......	Blood carrying waste substances passes through the kidneys. The poisonous wastes are removed from the blood and stored in the bladder. ✓

❑

Key ideas
Write 2 or 3 key words from each paragraph in this column

B5 Bacteria and viruses can cause infections that affect how the human body functions. For example AIDS and Hepatitis are diseases caused by viruses. Cholera and Typhoid are caused by bacteria. ❑

C1 Each organ system in a flowering plant also has a particular function. ❑

Write the name of each flowering plant organ in the correct box (see page 16).

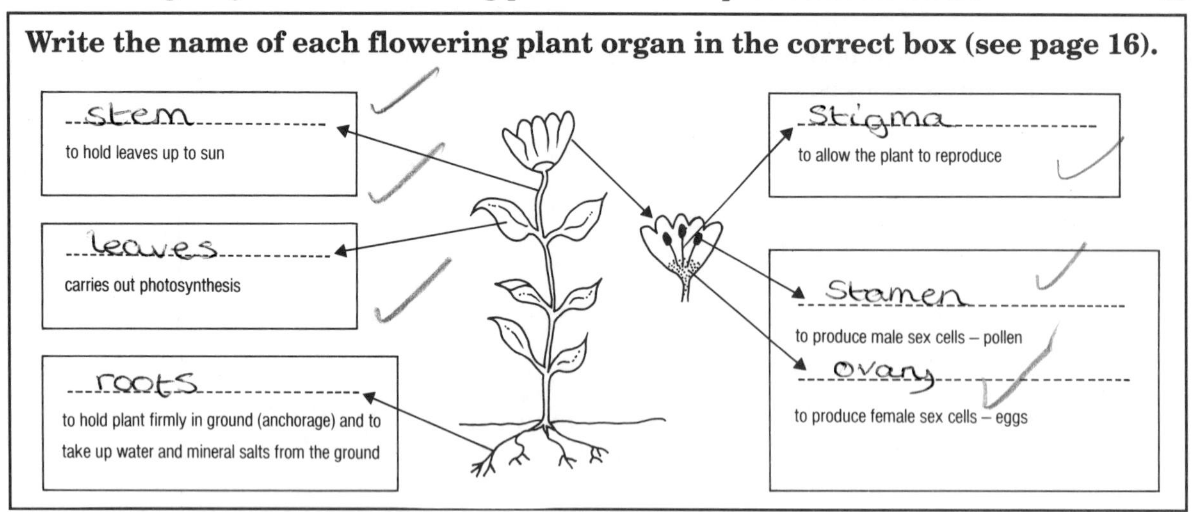

C2 Life cycles can be used to describe the reproduction of living things. ❑

Write down the correct names of the main stages in the life cycles shown below from the lists provided.

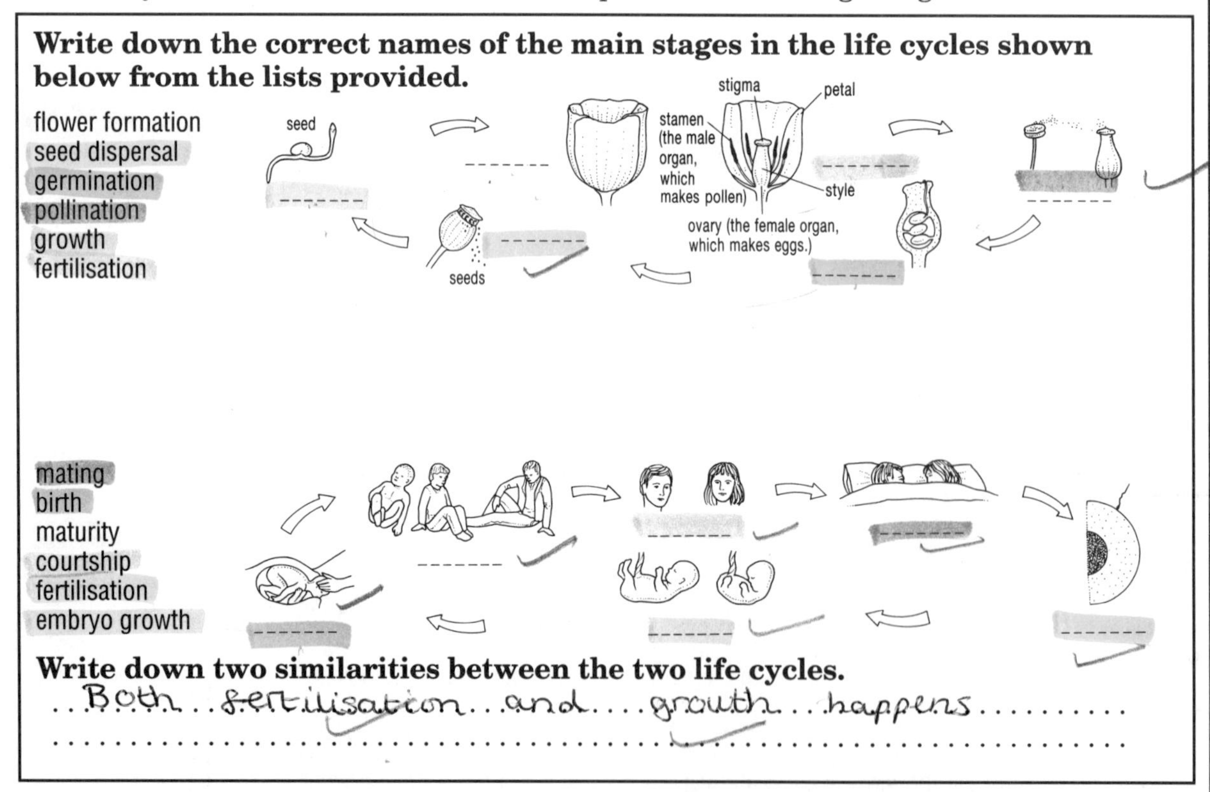

Write down two similarities between the two life cycles.

Both fertilisation and growth happens

D1 Grandparents, parents and children are members of three different generations. This can be shown in a family tree. ❑

Label the *oldest* generation and the *youngest* generation in this family tree.

Key ideas
Write 2 or 3 key words from each paragraph in this column

D2 Inherited features or characteristics are passed from one generation to the next in the form of genes. For example, there are genes for blood group and for eye colour. ❑

D3 There are a huge number of different kinds of living organisms. Scientists find it helpful to classify them in groups which share similar features. These features can be used to identify new organisms. ❑

Draw a line between the main plant or animal group and its description. Write 'animals' or 'plants' in the correct place.

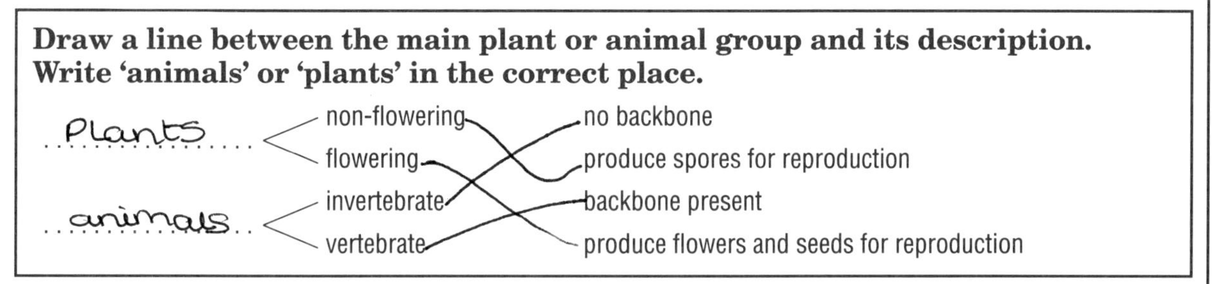

D4 The algae (seaweeds) and fungi (mushrooms and toadstools) used to be classified as non-flowering plants. Now they are put into groups of their own. ❑

Match each non-flowering plant with the correct description on the right.

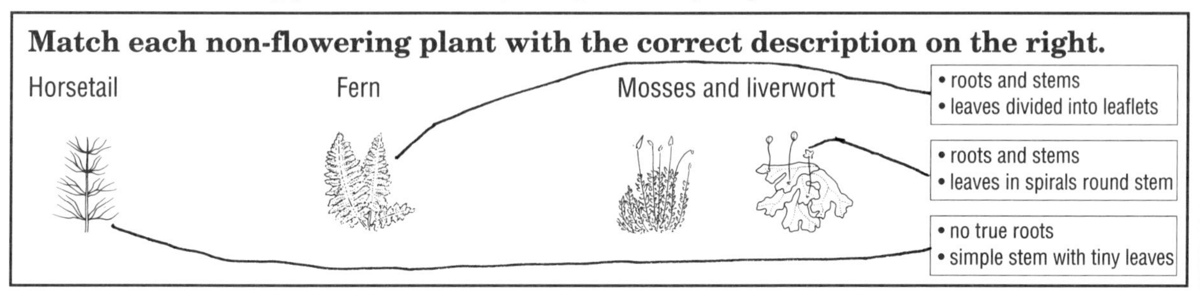

D5 Animals without backbones are known as invertebrates. ❑

Match each invertebrate with the correct description below.

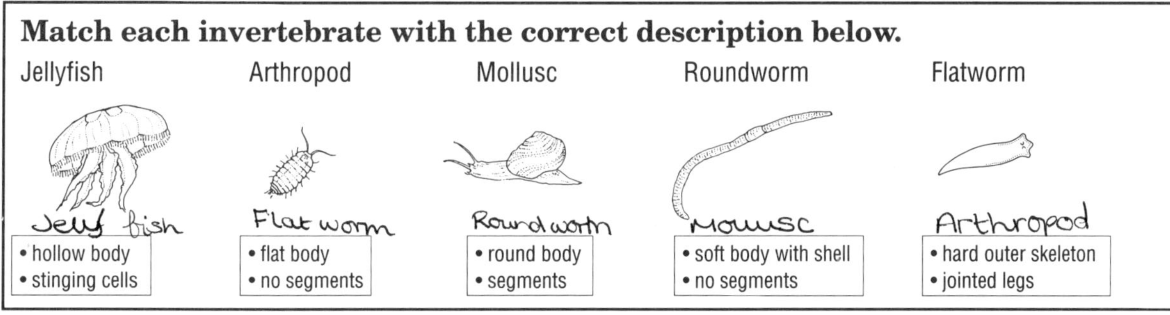

E1 Different organisms are found in different habitats. Environmental factors such as light, temperature or availability of water can vary between habitats. All the organisms in a habitat have adaptations that enable them to survive there. ❑

Complete the table.

Organism	Habitat	Environmental factor	Adaptation
cactus	desert	very hot, lack of water	able to store water
polar bear	arctic tundra	very cold	thick fur
shark	ocean	oxygen dissolved in water	gills
lemur	jungle	Humid, quite dark	very big eyes

E2 Waste material from human activity causes pollution which can affect the survival of living things. ❑

- smoke from chimneys – causes acid rain – kills trees
- car exhausts – release lead – damage to lungs and nervous system
- industrial chemicals – poisonous substances in food chain – poisons water life
- untreated human waste – reduces oxygen levels in water – kills water life

Underline or highlight key phrases.

A1 There are differences in structure between animal and plant cells. Each structure within a cell has a particular function. ❑

Write the names of the correct structures below each description of function. Label the three main structures of the animal cell.

Plant and animal cells	A typical plant and animal cell	Plant cells only
Controls cell activity. Contains the genes inherited from parents.	nucleus, vacuole, cytoplasm, cell wall, cell membrane, chloroplast plant cell	Outer covering of cell. Gives support and shape. ....
Controls the entry and exit of substances to and from the cell.		Contain the green substance chlorophyll which traps light energy in photosynthesis. ....
The jelly-like material within the cell. Lots of chemical reactions take place here.	animal cell	Contains a liquid called cell sap. Helps to support and control shape of cell. ....

A2 Pollination is the transfer of pollen from the stamen (the male organ of the flower) to the stigma (part of the female organ of the flower).
Fertilisation happens when the male sex cell in the pollen meets and joins with the egg to form a seed. ❑

B1 In aerobic respiration, energy is released from food in a useful form and in a controlled way inside cells. Normally, oxygen is required for the breakdown of the digested food to form carbon dioxide and water. This explains why we need to breathe in oxygen and why we breathe out lots of carbon dioxide and water vapour. This is called **gas exchange**. ❑

Write the words *aerobic respiration* and *gas exchange* in the correct places.

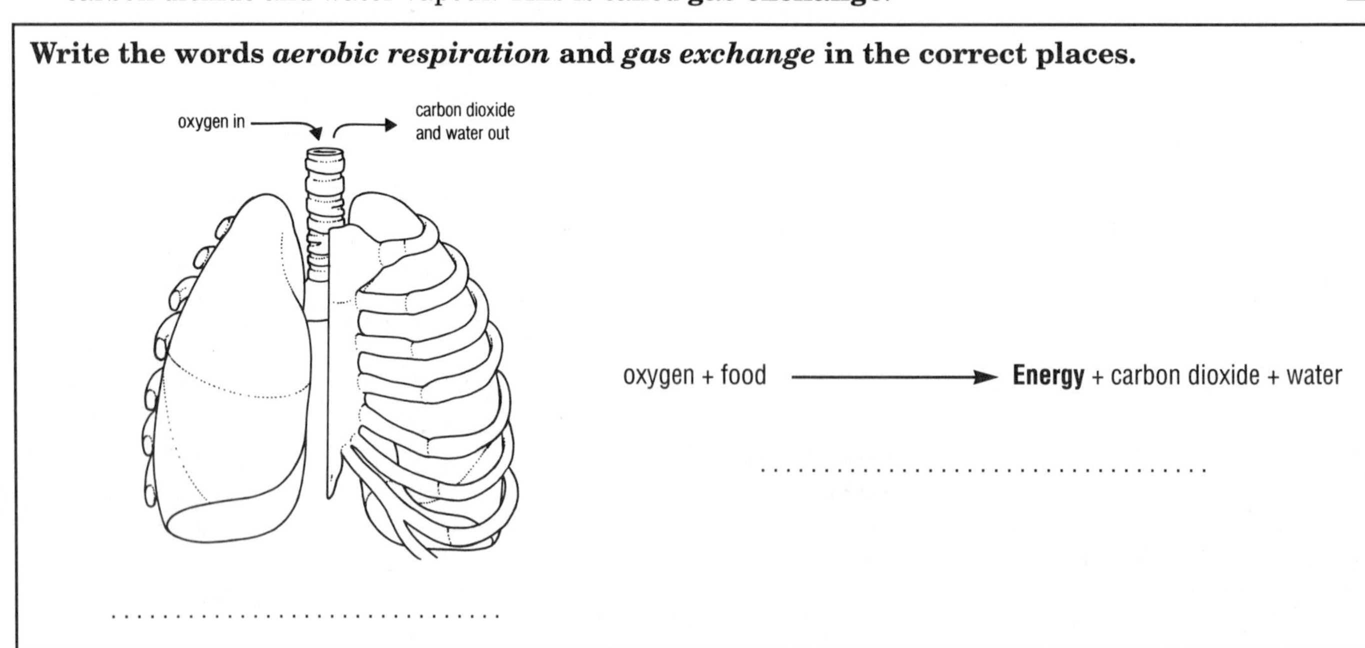

B2 Drugs such as alcohol, tobacco, glues, other solvents, cocaine and heroin can damage body organs and systems. They interfere with life processes. ❑

B3 Growing up involves physical, mental and emotional changes. ❑

Give the correct title (*physical changes* or *mental changes* or *emotional changes*) at the top of each description.

....	....	....
powers of reasoning increase problem-solving ability increases learns lots of new ideas and values	rapid growth reaches puberty improved co-ordination	understands rules and regulations develops responsible attitude to sex

C1 Green plants use energy from the sun to make their food (stored energy) from carbon dioxide and water. This process is known as photosynthesis.
Respiration, which happens in the cells of plants and animals, is the process by which glucose is broken down to produce carbon dioxide, water and energy. This energy is used for all living processes. ❑

Complete the following word equations and label them as either *photosynthesis* or *respiration*.

.... + ... —sunlight→ glucose + oxygen   process:

.... + oxygen ——→ .... + ... + energy   process:

C2 Materials such as carbon and nitrogen are cycled through an ecosystem to keep them at a steady level. Plants are very important in these processes. ❑

Label the diagrams *nitrogen cycle* or *carbon cycle*.

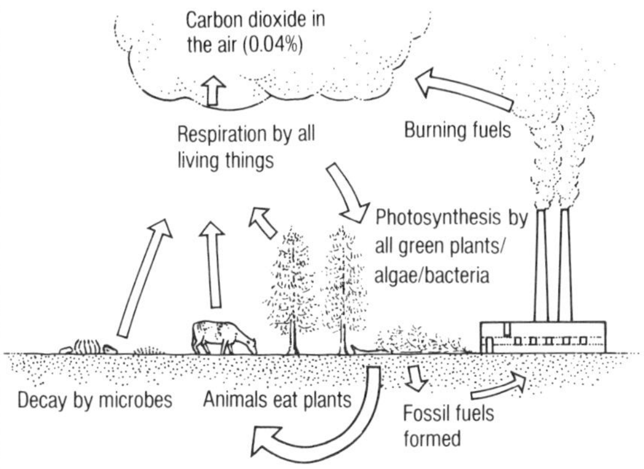

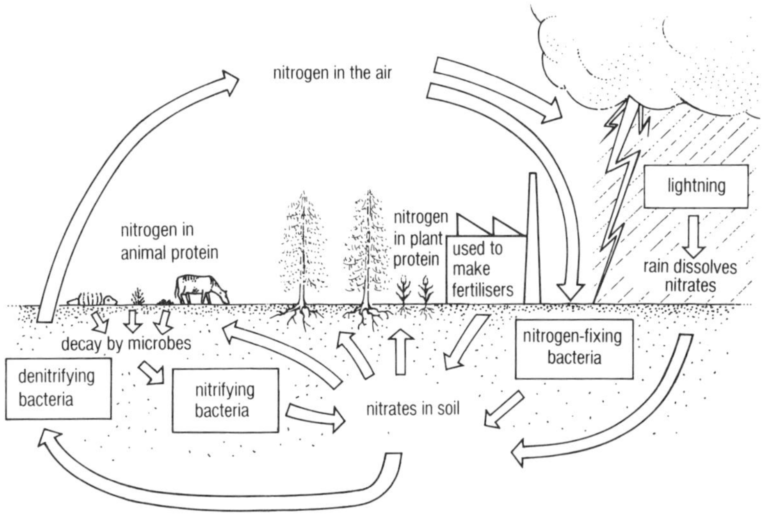

....

D1 Variation in living organisms can be due to inheritance (genetics) or to environmental factors. The weight of a new born baby is partly inherited and partly due to factors such as mother's diet and health and whether or not she smokes or drinks alcohol. ❑

Underline the variations which are mainly due to inheritance.

eye colour   temperature   blood group   right- or left-handed

E1 Organisms that live on land have adaptations for survival. ❑

Draw a line to match each problem of living on land with the associated adaptation shown by mammals.

Problem	Adaptation
reduce water loss	lungs for gas exchange
reproduce out of water	skeletal system
breathe air	internal fertilisation and development
support themselves	waterproof covering

E2 Organisms which are adapted for one kind of habitat will not be found in large numbers in a habitat which is very different. ❑

E3 The numbers in a population are controlled by competition for resources, by the forces of nature (for example good or bad weather), by disease and by the activity of predators. ❑

E4 Organisms are linked in food chains, and any change in one population will therefore affect other populations. ❑

Complete the sentence to describe what you expect to happen.

Sun → Grass → Rabbit → Buzzard

If the rabbit population is reduced by disease the amount of grass will ....

and the number of buzzards will ....

E5 There are many interconnected food chains in an ecosystem. They build into a **food web**. ❑

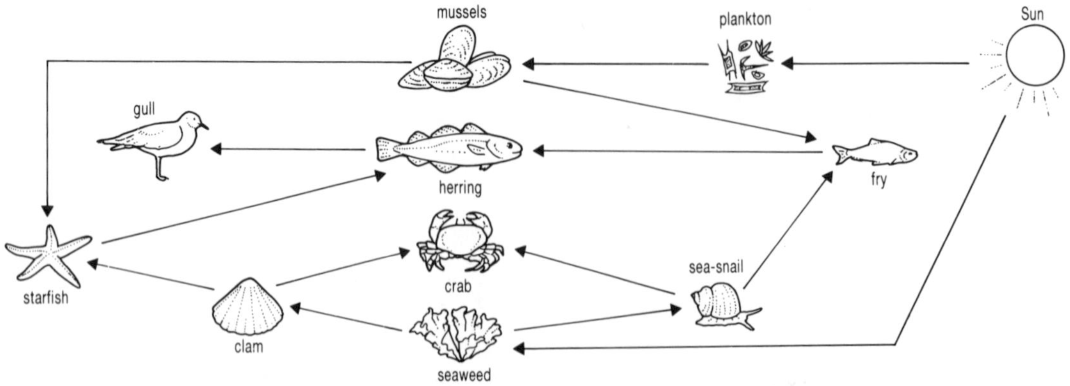

mussels
plankton
Sun
gull
herring
fry
starfish
crab
sea-snail
clam
seaweed

Write down two food chains from this food web.

....

....

E6 Toxic materials can be concentrated at the end of a food chain. ❑

Underline or highlight key phrases.

A1 Organs and organ systems are made up of tissues. Each different tissue contains cells which are adapted to carry out the same specialised function. For example, muscle tissue in the heart can contract. ❑

A2 There are lots of different types of cells in plants and animals. Each cell type has a structure that helps it to carry out its function. ❑

Human nerve cell or neurone

very long, covered in insulation, for rapid movement of nerve impulses over long distances

Plant leaf palisade cell

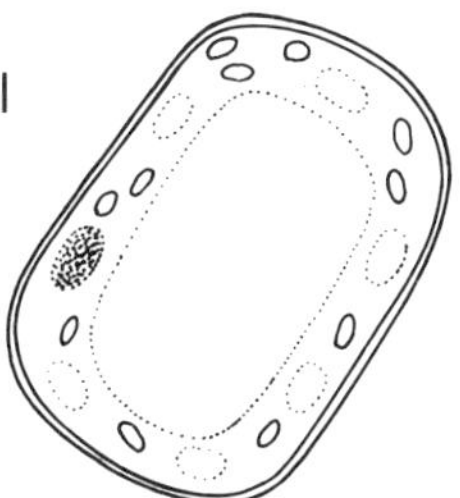

long cylinder-shaped cells with many chloroplasts to trap light for photosynthesis

Human red blood cell

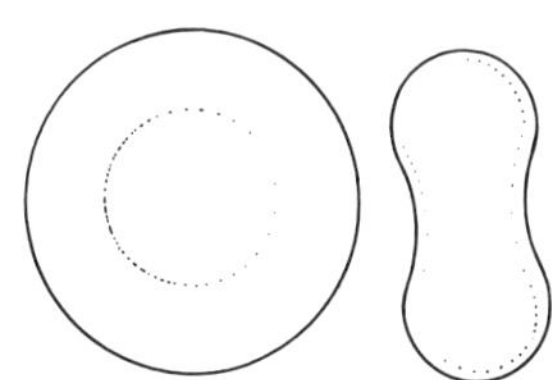

dumb-bell shape gives a big surface area to collect oxygen

Root hair cells

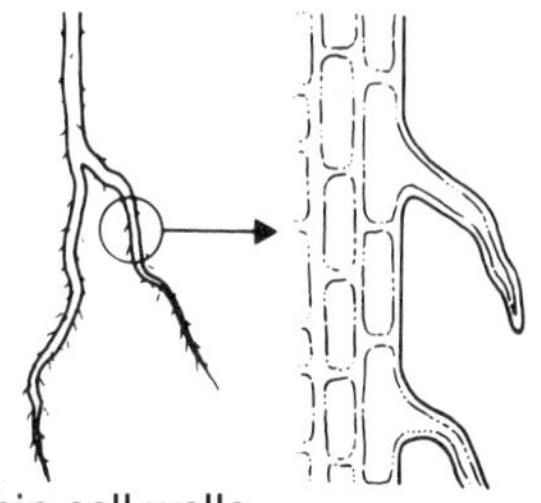

long cells with thin cell walls to absorb water from the soil ❑

Name two other examples of cell adaptations – one from a plant and one from an animal.

B1 Feeding or nutrition results in the chemical breakdown of food by enzymes to form soluble molecules.
- Each food type – carbohydrate, fat or protein – is broken down by a particular enzyme that works best at body temperature, 37 °C.
- Undigested food passes through the large intestine, where excess water is absorbed into the bloodstream, before being egested at the anus.
- Digested food is absorbed into blood vessels lining the small intestine, and pumped round the body to all cells by the heart. This food may be used for growth, repair, or as an energy source for the process of cellular respiration. ❑

B2 The structure of the lungs helps the process of gas exchange. Millions of tiny air sacs provide a large surface area for gas exchange. Air sacs:
- are very thin and moist to help speed up gas exchange
- have a rich blood supply to absorb oxygen **from** the lungs and to pass carbon dioxide and water back **into** the lungs. ❑

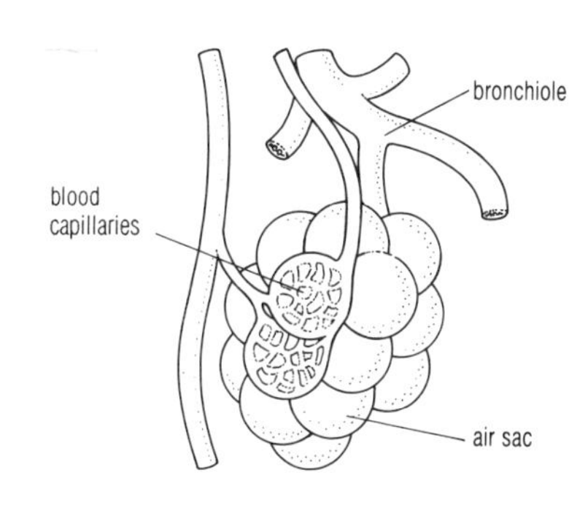

Write the four important features of lungs here.

B3 When you smoke you damage the structure of your lungs. This makes gas exchange less efficient. Smoking may lead to lung cancer.

Put these statements in the order in which they happen to a smoker.

A smoker's cough develops

B chemicals in smoke damage tiny hairs (cilia) lining windpipe

C coughing damages air sacs

D mucus and bacteria not removed from lungs

E severe breathlessness develops

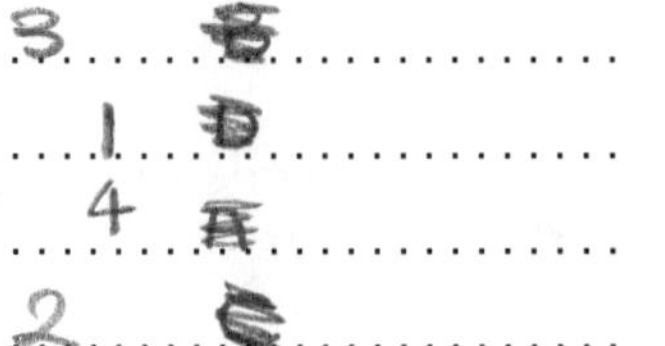

B4 The action of rib muscles and the diaphragm inflates and deflates the lungs:

- rib muscles contract – ribs move out
- diaphragm contracts and pulls down

Write the labels *inflated* and *deflated* under the correct diagram.

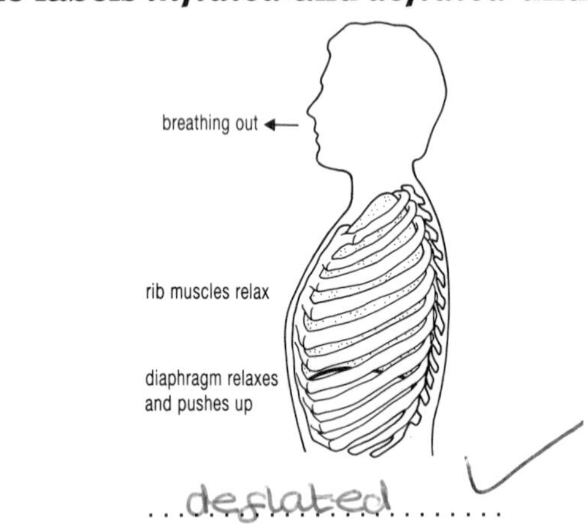

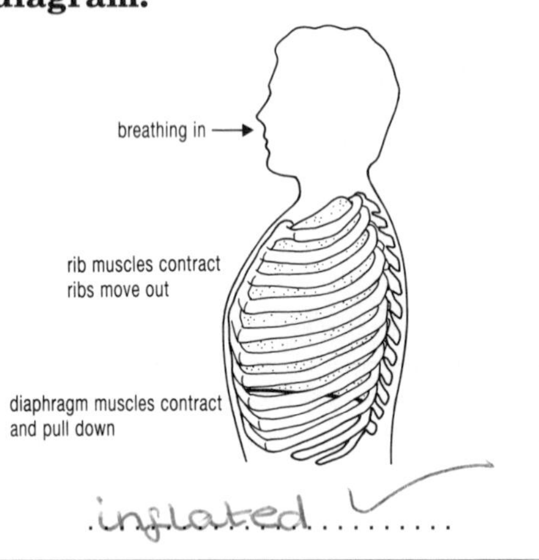

B5 The kidneys filter poisonous substances (such as urea) and waste substances (such as extra salts) from the blood. These are passed to the bladder for storage before being excreted as urine. The kidneys also keep the amount of water in the blood in balance by controlling the volume of water excreted in the urine. For example, after you have eaten a lot of salty or sweet food, you excrete a small volume of concentrated urine.

Underline one word from each pair to make this sentence correct.

After drinking a lot of water, you will excrete a 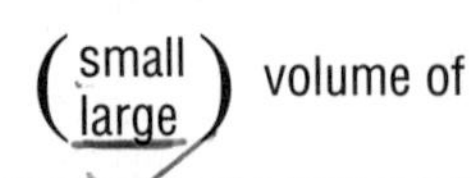 (small / large) volume of  (dilute / concentrated) urine.

B6 Movement depends on a pair of muscles at each joint. Energy is used for one muscle to contract as the other relaxes.

Label the contracting muscle C and the relaxing muscle R in the second diagram below.

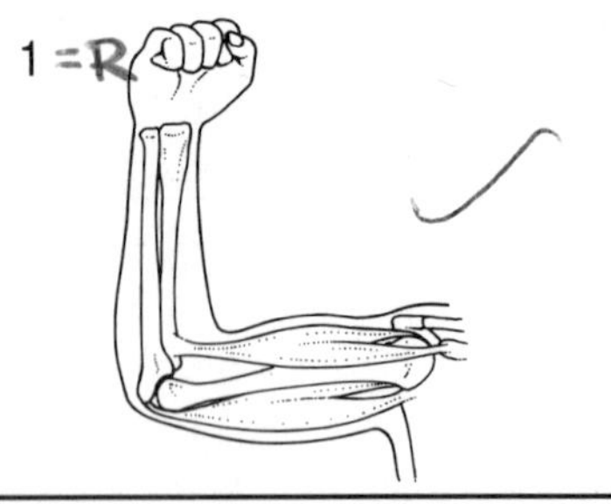

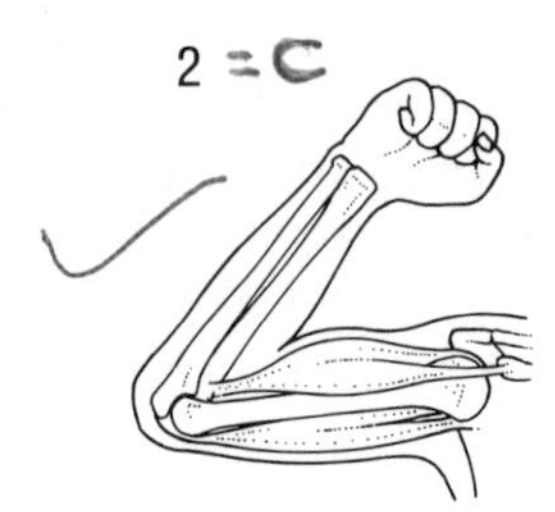

B7 **Put the following events of sexual reproduction in humans in the correct order, numbered 1–7.**

Number	Events in sexual reproduction	Number	Events in sexual reproduction
3	**A** sexual intercourse takes place	2	**E** egg reaches the fallopian tube (egg tube)
6	**B** fertilised egg becomes attached to wall of uterus	4	**F** sperm swim from vagina through uterus to fallopian tube
1	**C** egg released from ovary (ovulation)		
5	**D** one sperm fertilises the egg	7	**G** embryo begins to grow and develop

B8 Food and oxygen are passed from the mother's blood to the developing baby's blood across the placenta. Waste matter from the baby passes in the opposite direction. The umbilical cord joins the baby to the placenta. ❑

Label the placenta and umbilical cord.

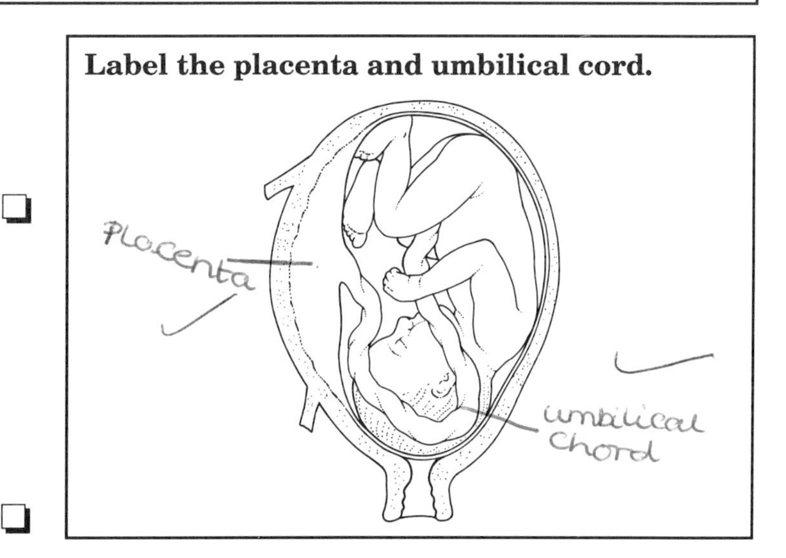

B9 All animals sense changes (stimuli) in their environment and respond to these. This is called sensitivity or behaviour. Receptors change the stimulus into electrical messages that are sent along nerves (neurones) to the brain. The brain controls any responses by sending messages along neurones to muscles. ❑

Complete the table.

Sense organ or receptor	Sense	Stimulus
skin	touch	pressure
eye	See	light
ear	hearing	noises
tongue	taste	chemicals
nose	Smell	chemicals
inner ear	balance	gravity

C1 The green colouring in plants is called chlorophyll. This substance traps light energy from the sun and uses it to change carbon dioxide and water into sugar (glucose) and oxygen. Simple sugars can be changed into a variety of substances for growth and repair, or stored as starch, or used in respiration to provide energy. Plants must respire all day and all night to stay alive. ❑

Write the words *stored, growth and repair* and *aerobic respiration* in the correct places.

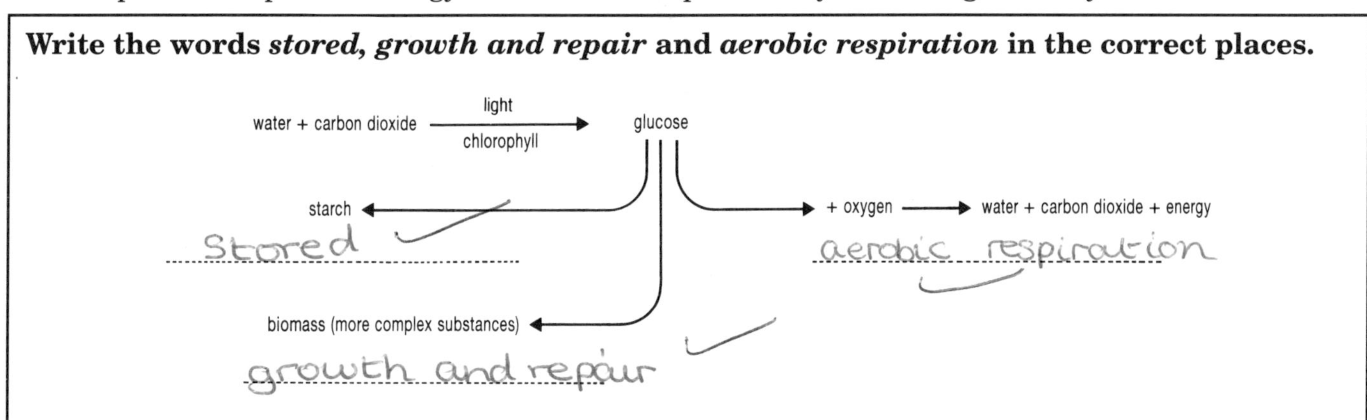

C2 Plants require certain minerals for healthy growth. In particular, they need minerals containing the elements nitrogen, potassium and phosphorus. ❑

D1 Improved varieties of crop plants and domesticated animals can be produced by selective breeding. This means that the parents are carefully chosen because they have desirable characteristics. The breeder hopes that their offspring will inherit these characteristics. For example, dairy cattle are bred to produce more milk, and several wild strains of wheat have been crossed to produce modern varieties with higher yields and improved resistance to disease. ❑

D2 Organisms of the same species show variation. The variation can be due to inherited characteristics or the effect of the environment on inherited characteristics. In humans eye colour is inherited, whereas skin colour depends on inherited factors and the effect of sunlight. ❑

Circle the characteristics in the list below that are the result of inheritance, underline the characteristics that depend on inheritance and environmental factors.

ability to roll your tongue

coat colour in gerbils

height in humans

variation in size of Granny Smith apples

weight in humans

E1 The growth and decline of any population depends on the availability of environmental resources. For example, when a population of bacteria grows in a culture, the numbers increase very rapidly at first when there is a plentiful supply of oxygen and food. The numbers increase more slowly and eventually level off as these resources get used up. ❑

Complete the graph to show what would happen to the population when all the food had been used up.

Number of bacteria

Time

A typical growth curve

E2 A food chain shows the path of energy transfer through organisms in an ecosystem, but it does not give a measure of the amount of energy transferred at each stage of the chain. Pyramids of number or pyramids of biomass can be used to do this. ❑

Identify the diagrams with the titles *biomass* or *number*.

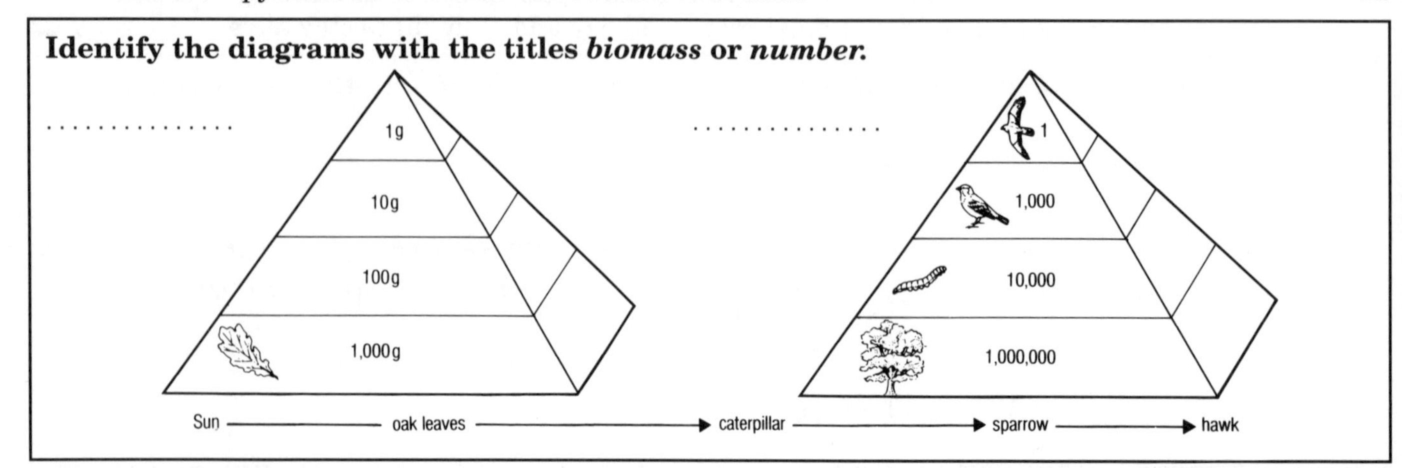

ATTAINMENT TARGET 3

Materials and their Properties

Big ideas

A There are different types of materials with different properties. Properties can be explained by thinking about particles.

B Materials can be changed by physical actions, geological actions or chemical reactions.

C Chemical reactions show patterns.

Key ideas
Write 2 or 3 key words from each paragraph in this column

A1 A material is a useful substance. ❑

A2 Some materials are natural (found in or on the Earth).
Some materials are synthetic (made by people). ❑

Natural

Synthetic

Draw two more natural materials and two more synthetic materials.

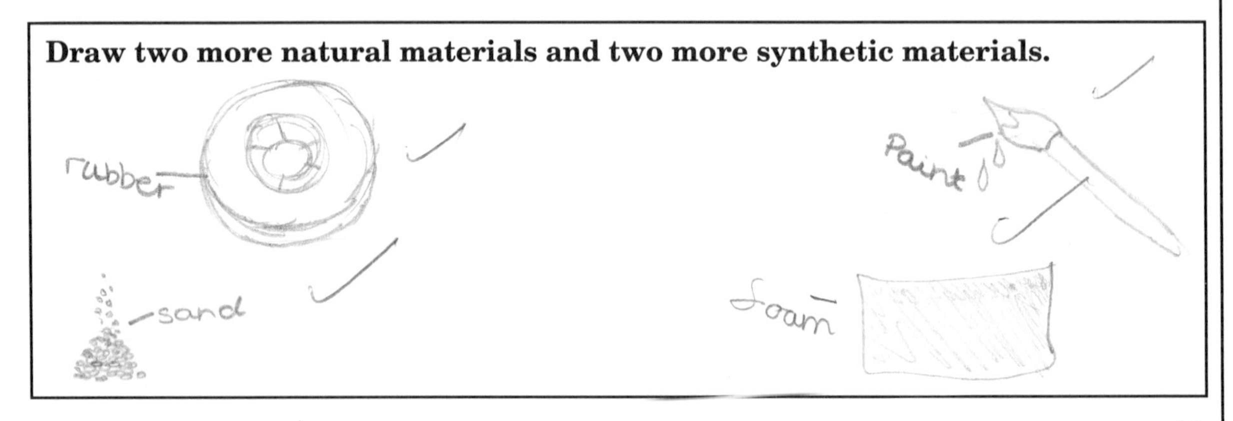

A3 Materials can be put into groups. ❑

Put these materials into groups. Circle each group in a different colour. For example, the *ceramics* are already ringed in black.

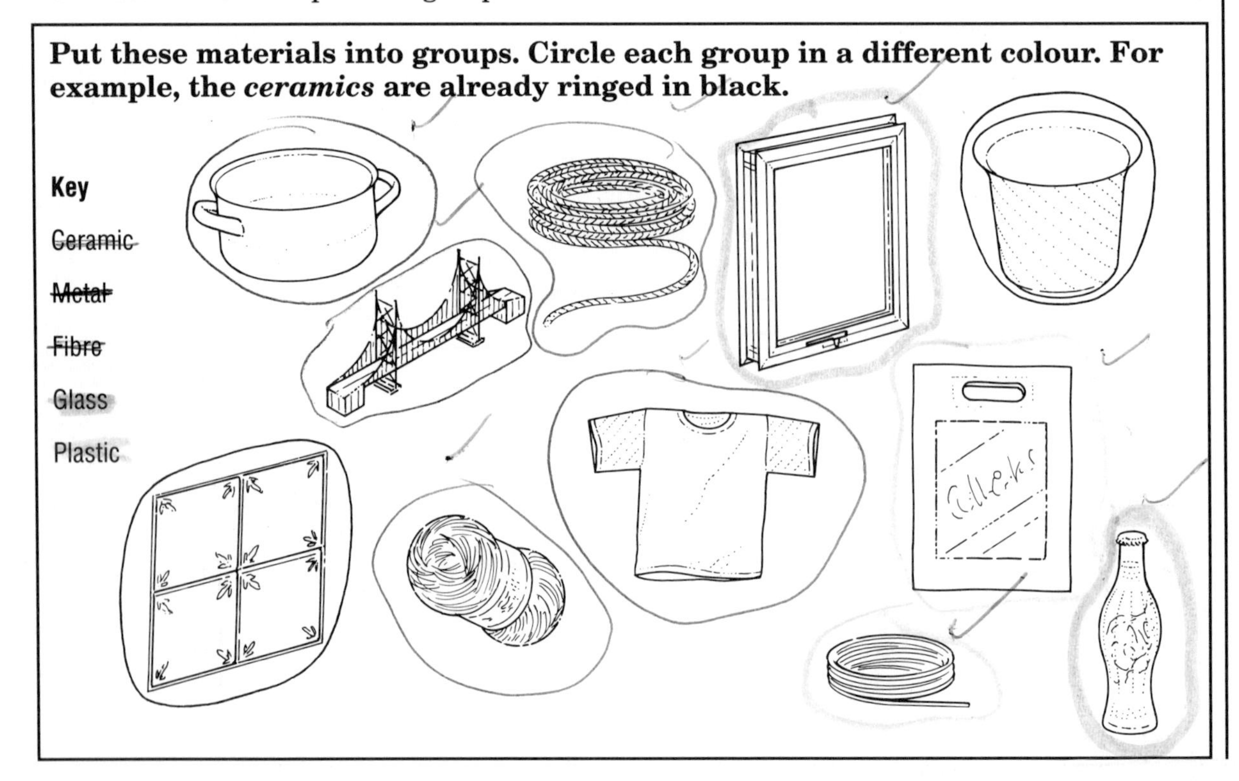

Key ideas
Write 2 or 3 key words from each paragraph in this column

A4 The use of a material depends on its properties. Glass is useful because it is:
- transparent (see-through)
- waterproof (keeps water out)

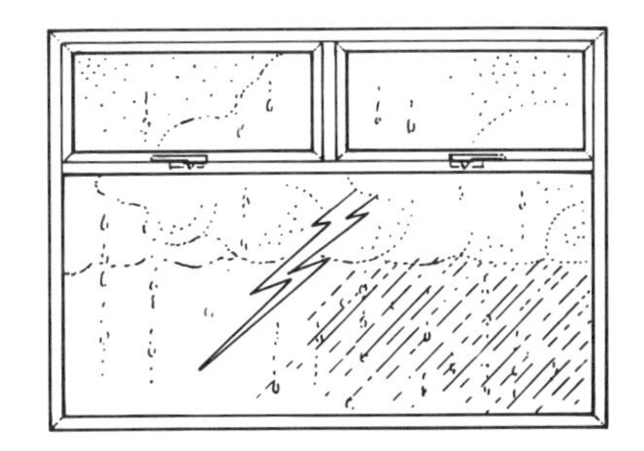

Metal is useful because it is:
- an electrical conductor (lets electricity through)
- a thermal conductor (lets heat through)
- flexible (easy to bend)

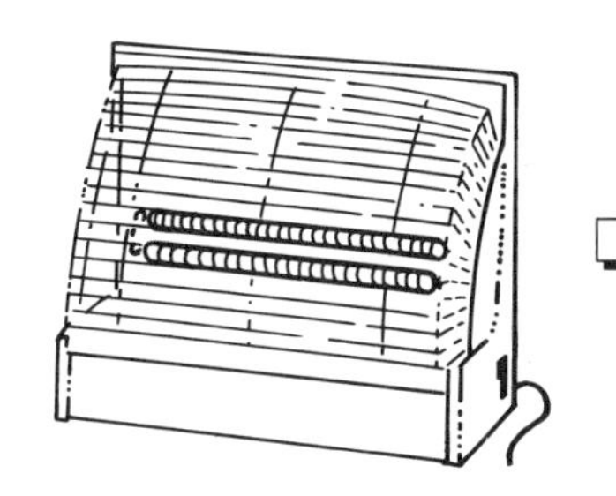

Complete the table with your own examples.

Material	Important property
glass	transparent
metal	hard
wood	strong
stone	rough
foam	flexible
plastic	waterproof
rubber	dense
balsa wood	breakable

B1 Materials can be changed physically or chemically. Some physical changes can be reversed easily, chemical reactions cannot.

Circle the illustrations which show a physical change which can be reversed easily.

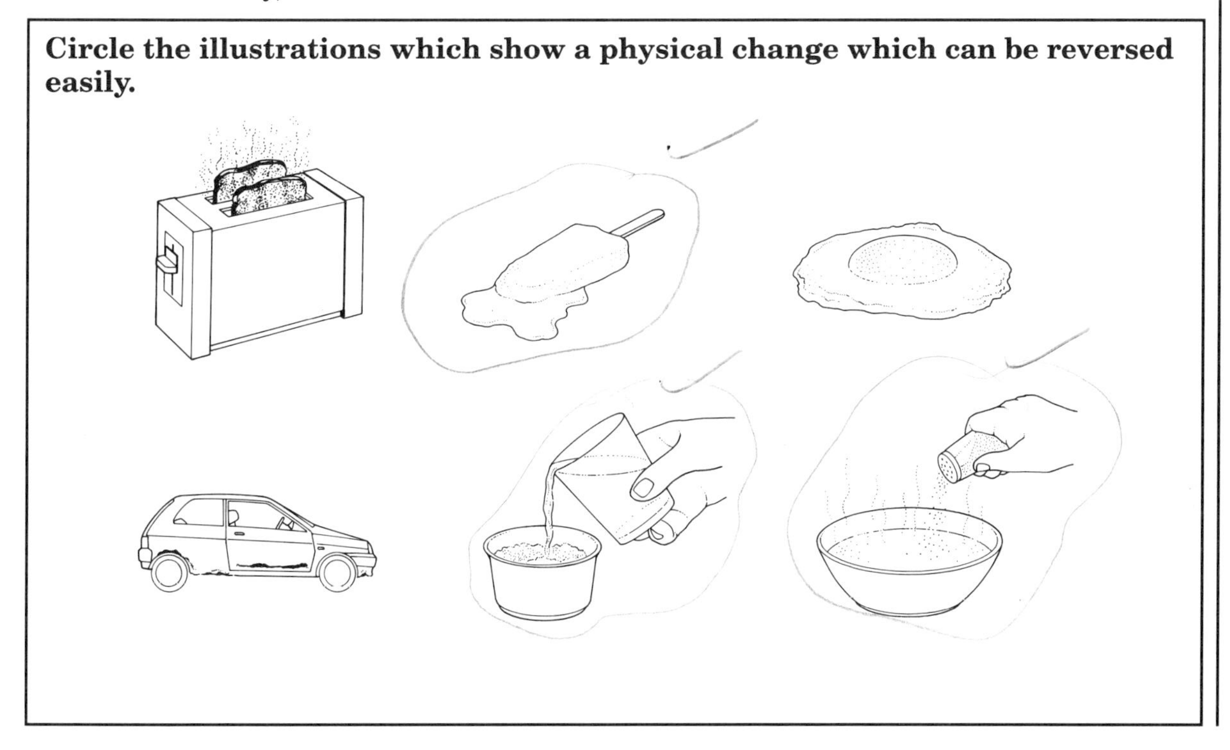

Key ideas
Write 2 or 3 key words from each paragraph in this column

A1 Materials come from a variety of sources. Gases like oxygen and neon come from the air; liquid fuels like petrol and diesel come from oil; solids like metals come from rock. Some materials, like wool and sugar, come from living things. ❑

A2 A material can be a solid, a liquid or a gas. For example, water can be solid ice, liquid water or a gas (water vapour). Therefore, water can have different states. ❑

Write five examples in each column.

Solid	Liquid	Gas
brick	water	oxygen
wood	orange juice	helium
Plastic	tea	carbondioxide
glass	coffee	nitrogen
metal	coke	argon

A3 Materials can be classified as solid, liquid or gas by looking at simple properties. For example, in a car:

- the metal keeps its shape
- the petrol fits in the tank
- the air fills up the tyre ❑

Solid (metal):
- does not flow
- keeps its shape

Liquid (petrol):
- flows easily
- changes shape

Gas (air):
- flows very easily
- changes shape to fill container ❑

A4 The state of a material depends on the temperature. Different materials change state at different temperatures. When the state of the material changes, some of the properties of the material will also change. ❑

Add these labels to the diagram.

evaporation

condensation

freezing

melting

A5 Mixtures can be separated using several different techniques. ❑

Draw a line to show which technique would be the best way to separate each substance from the mixture.

sand from a mixture of sand and water	sieving
pebbles from a mixture of pebbles and sand	filtration
water from a mixture of salt and water	evaporation
salt from a mixture of salt and water	condensation

B1 Substances can be changed into other substances. This change is called a chemical reaction. Almost all materials, including those in your body, have been made by chemical reactions. You can spot a chemical reaction because it often involves a transfer of energy to or from the surroundings.

- burning (**combustion**)
  e.g. burning wood to heat the house
- strong heating (**decomposition**)
  e.g. heating wood to make charcoal ❑

Key ideas
Write 2 or 3 key words from each paragraph in this column

B2 Many raw materials need to be changed into more useful ones by a chemical reaction.
- An ore is **smelted** to make metal (like copper).
- An acid is **neutralised** to make a salt (like ammonium nitrate, a fertiliser).
- A sugar is **fermented** to make alcohol.
- An element is **burned** in oxygen to make an oxide. ❑

B3 Substances can undergo physical changes and chemical changes. Some physical changes such as dissolving, freezing or evaporating, are reversible. Other changes, including a chemical change such as burning, are not reversible. ❑

Write down two examples of changes and say whether you think they will be reversible or irreversible.

water freezing = reversable

melted candle wax = irreversable

B4 A **fuel** is a substance that burns in oxygen to release heat energy. Waste gases are usually produced too and these may cause damage to the environment. Common fuels are natural gas, petrol and coal. ❑

B5 Materials can be changed by the weather. Heat, rain, cold and wind all change materials. For example, building materials change colour because of pollution and acid rain. ❑
Some stone will even dissolve slowly in rain. It will lose its detailed features. Small cracks in rock and stone let water in. This can freeze. The ice expands to open the crack. Eventually, the rock or stone may break and crumble. ❑
Weathering can wear down mountains. ❑

B6 Weathering can lead to erosion of rock. The pieces of rock are then carried far away by wind and water and ice. Similar pieces may be dropped together in a layer (or sediment). ❑

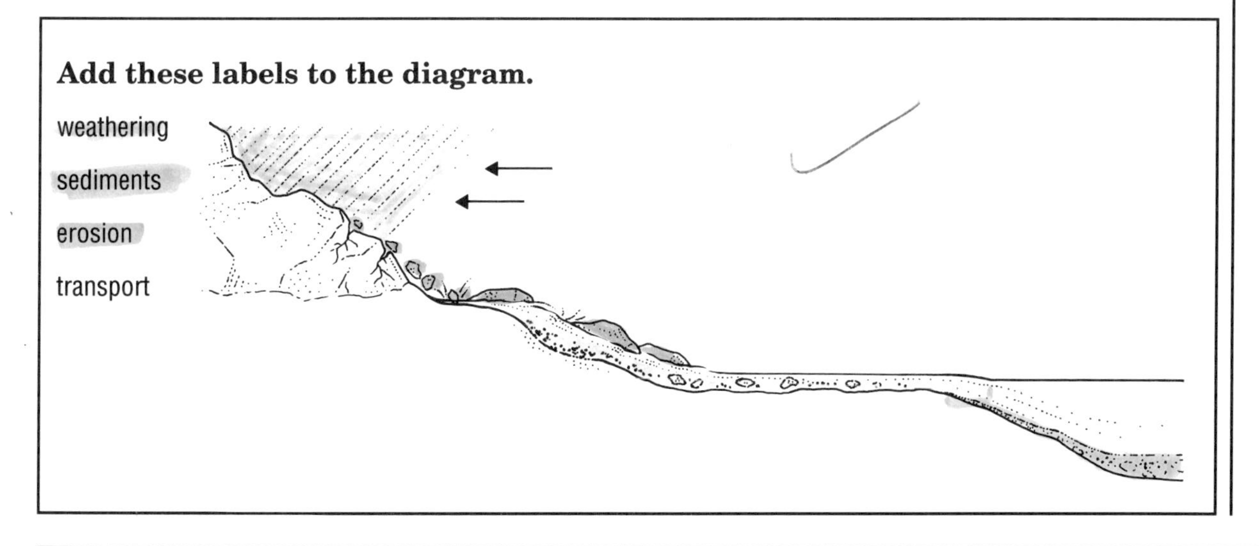

Key ideas
Write 2 or 3 key words from each paragraph in this column

A1 Everything is built from elements. All the elements are listed in the periodic table, with metals on the left-hand side and non-metals on the right. ❑

Mark the metals in the table. Also identify by name any groups of elements that you can recall.

alkaline metals

TRANSITION METALS

HALOGENS.

							H										He
Li	Be											B	C	N	O	F	Ne
Na	Mg											Al	Si	P	S	Cl	Ar
K	Ca	Sc	Ti	V	Cr	Mn	Fe	Co	Ni	Cu	Zn	Ga	Ge	As	Se	Br	Kr
Rb	Sr	Y	Zr	Nb	Mo	Tc	Ru	Rh	Pd	Ag	Cd	In	Sn	Sb	Te	I	Xe
Cs	Ba	La	Hf	Ta	W	Re	Os	Ir	Pt	Au	Hg	Tl	Pb	Bi	Po	At	Rn
Fr	Ra	Ac	Ku	Ha													

– metals
– noble gases

A2 Elements join together to form compounds.
- Copper chlor**ide** is made from copper and chlorine **only**.
- Copper chlor**ate** is made from copper, chlorine **and** oxygen. ❑

A3 Elements and/or compounds can be mixed to form mixtures. ❑

Identify the sets by writing *elements, compounds* or *mixtures* underneath.

Silver nitrate	Black ink	Copper
Chalk	Crude oil	Oxygen
Water	Coal	Sodium
compounds ✓	mixtures ✓	elements ✓

A4 An important scientific hypothesis is that all substances are made from tiny particles, called atoms. For example, an element is a substance that contains **one kind** of atom. ❑

A5 Most of the elements in the periodic table are metals. ❑

Underline the properties of metals.

- shiny ✓
- dull
- conduct heat ✓
- do not conduct electricity
- change state when hammered
- gas at room temperature
- can be drawn out to make wire ✓

Key ideas
Write 2 or 3 key words from each paragraph in this column

A6 Muddy water and seawater are examples of mixtures. The substances in a mixture can be separated easily: they are not joined. For example:

To separate mud from water:
1 decant the water
2 filter the water.

To separate salt from seawater:
3 evaporate the water
4 crystallise the salt.

Identify each separation method with its number (1–4) .

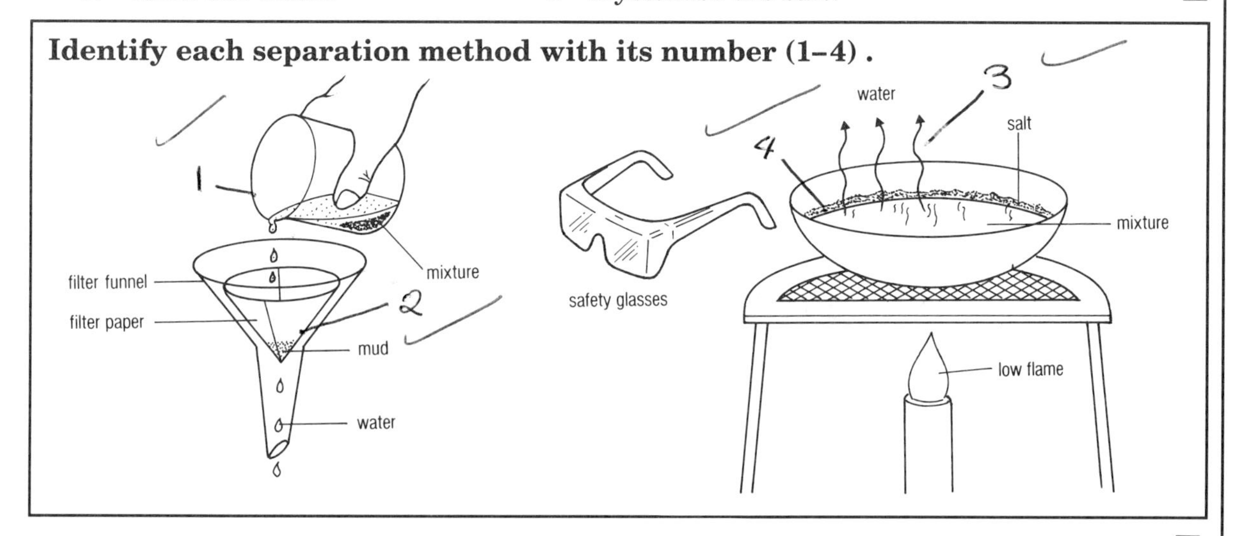

A7 Crude oil is a mixture of chemicals. It can be separated by fractional distillation.

Complete the diagram by writing the labels in the correct boxes. Add your own labels to the empty boxes if you can.

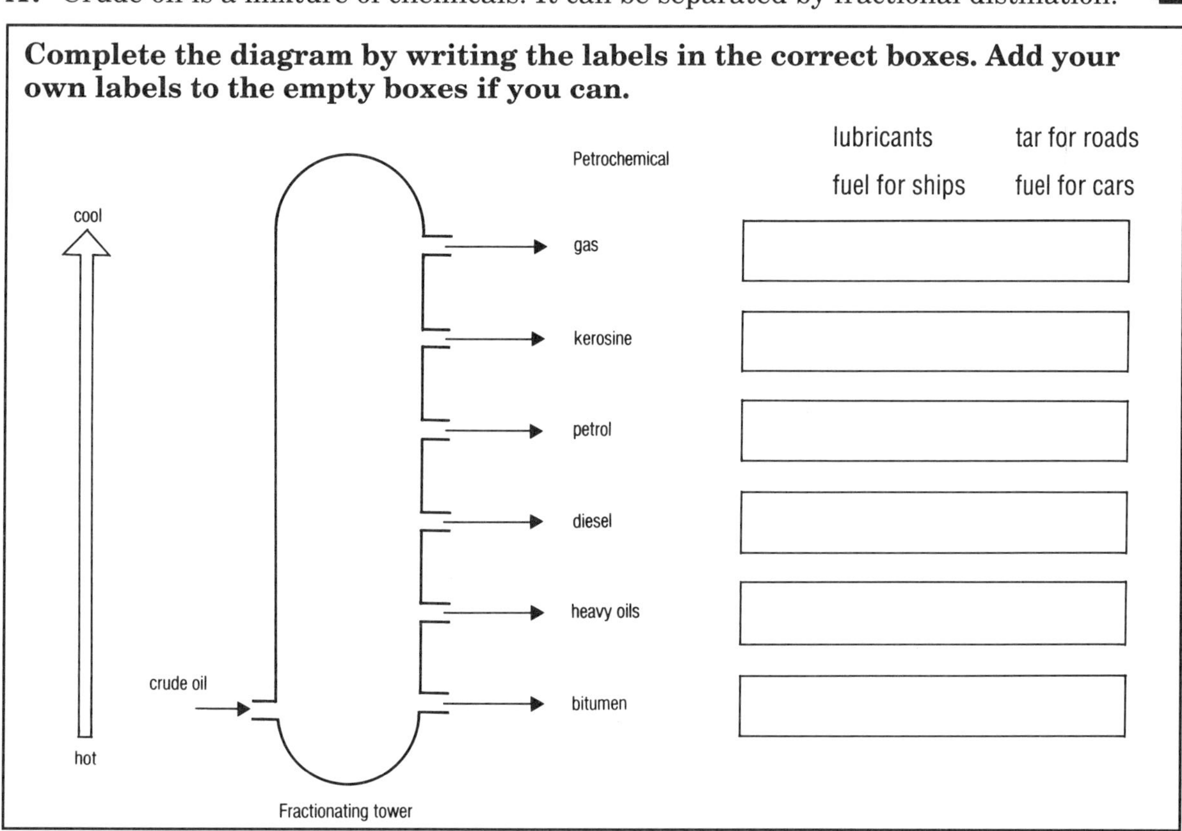

B1 Some physical changes occur naturally in the water cycle.

Match the correct halves together to describe the water cycle. (See page 32 for a diagram.)

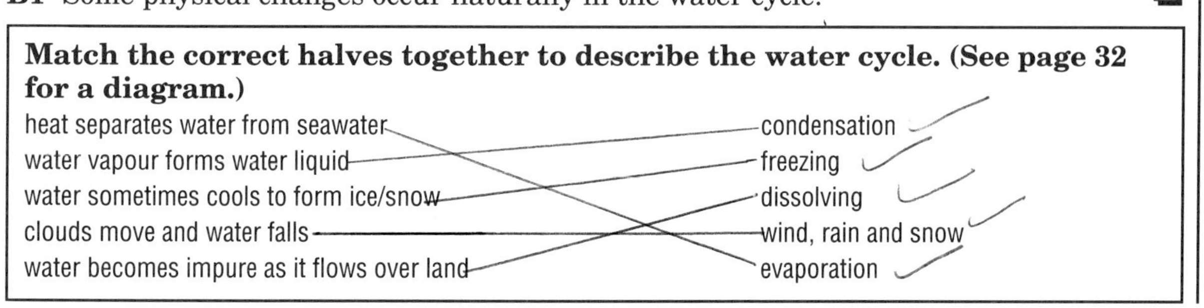

Key ideas
Write 2 or 3 key words from each paragraph in this column

B2 Water is a compound. It is made by joining hydrogen and oxygen.

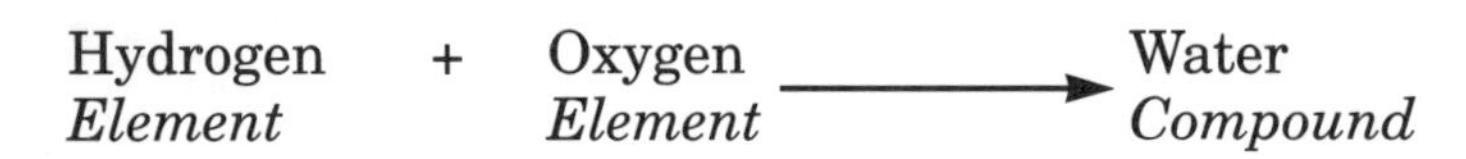

Water is often made when a substance that contains hydrogen atoms is burned. Another common product of burning is carbon dioxide.

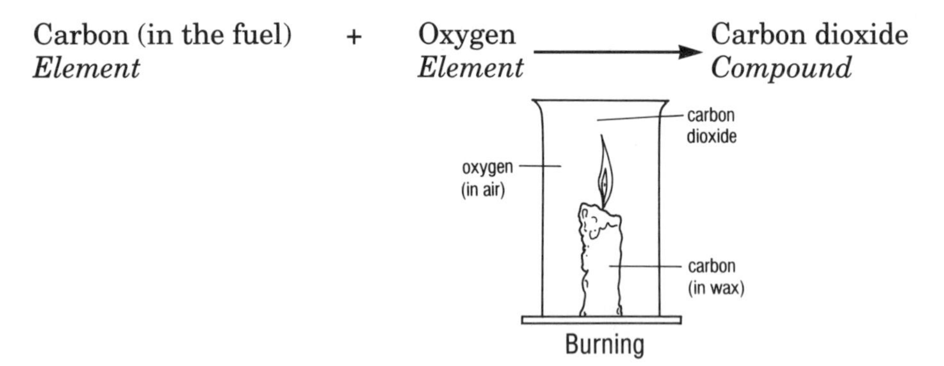

❑

B3 Burning is when an element joins with oxygen. So is rusting.

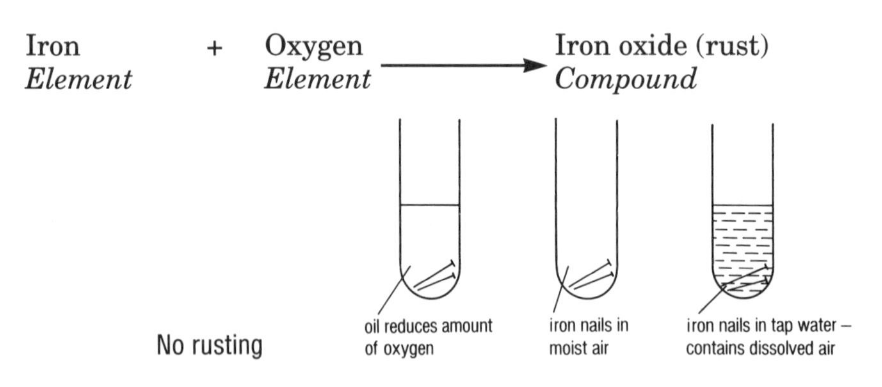

No rusting | Rusting ❑

B4 Rusting and burning can be slowed down by reducing the amount of oxygen available. ❑

Underline in one colour the substances which are used in anti-rusting methods. In another colour underline those used in anti-burning methods. Add a colour key.

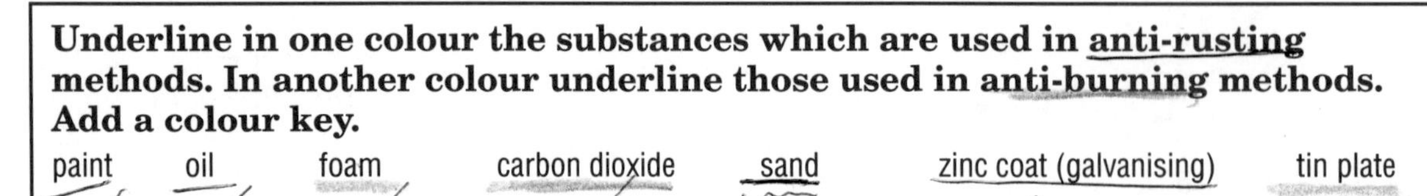

C1 Acids are also compounds. Acids can be cancelled out by alkalis and become neutral. (Neutral is neither acid nor alkali.) ❑

C2 Acids and alkalis can be detected by indicators. An indicator is a chemical that changes colour when the pH changes. Universal indicator has a range of colours. ❑

Add the universal indicator colours to the panel in the diagram.

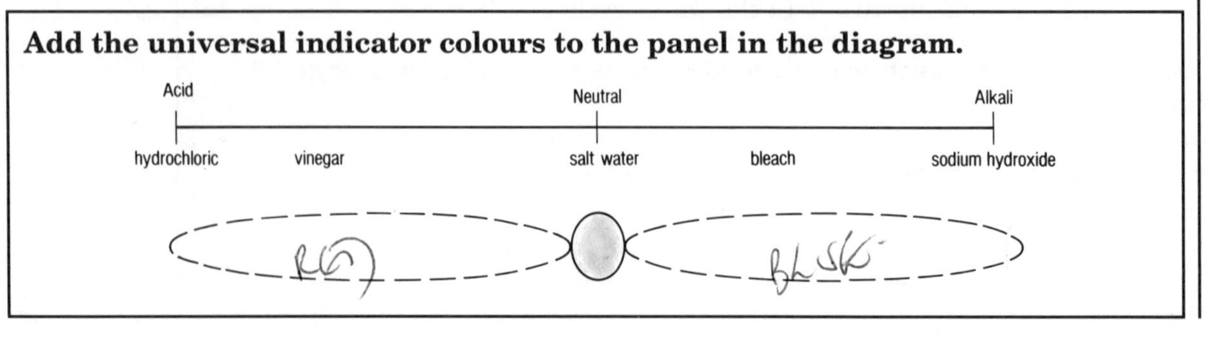

Underline or highlight key phrases.

A1 Metals are very useful materials. They are shiny, easily shaped and good conductors of heat and electricity. ❑

Use a periodic table to complete the table with ticks or crosses as appropriate.

Element	Metal?	Does it conduct electricity?
S		
Hg		
Na		
O		
I		
Cu		

A2 One exception to the above pattern is carbon: it is a non-metal, but it conducts electricity. ❑

A3 Particles are arranged differently in solids, liquids and gases. Ice keeps its shape because the particles are held together strongly. Water changes shape because the particles are not held together as strongly so they can move around more easily. ❑

Use these ideas to work out why the particles in water vapour can spread all round the room. Illustrate your ideas by drawing the arrangement of particles in a solid, a liquid and a gas.

In a solid: slow moving, small spaces   In a liquid: faster moving, bigger spaces   In a gas: very fast moving, very big spaces

B1 When you heat a substance, its particles move more. The spaces between particles increase. The substance will then get bigger (expand). When it cools, the spaces decrease and the substance will then get smaller (contract). The size of the particles does not change. ❑

B2 In physical changes and in chemical reactions the total mass of the substances is conserved. ❑

Underline the correct end to the sentence below.

Conserved means stays the same/gets bigger/gets smaller.

B3 If a substance is heated in oxygen, it may burn. Burning is a chemical reaction with oxygen, and compounds called oxides are formed. ❑

Complete these word equations. (Note that ethanol contains carbon, oxygen and hydrogen atoms.)

carbon + oxygen ⟶ carbon dioxide

copper + oxygen ⟶ ?

? + ? ⟶ sulphur dioxide

magnesium + oxygen ⟶ magnesium oxide

hydrogen + oxygen ⟶ ?

ethanol + oxygen ⟶ ? + ?

B4 A reaction where a substance joins to oxygen is called **oxidation**. Oxidation of metals causes corrosion. Oxidation of food causes spoiling. The opposite reaction, where a substance loses oxygen, is called **reduction**. ☐

B5 Oxidation and reduction often cause atoms to become charged. Charged atoms are called ions.
- Metal ions are always positive.
- Non-metal ions are always negative. ☐

Circle the ions in this list.

Ne Na^+ Mg Cl C Cl^- Na O^{2-} Mg^{2+} Al

B6 Reactions which give out heat energy (like combustion) are called **exothermic** reactions. Those which take heat in are called **endothermic** reactions. ☐

Label the drawings as either *exothermic* or *endothermic* reactions. The room temperature is 20 °C.

Freezer pack 1 °C
....

Burning coal 400 °C
....

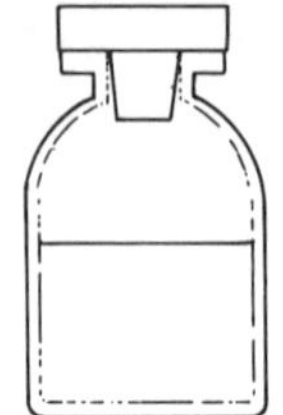
Neutralising acid 22 °C
....

Animal 37 °C
....

B7 Rocks have been made by chemical reactions and physical changes. There are three main types of rocks: **igneous**, **metamorphic** and **sedimentary**. The diagram illustrates how each type was formed. These processes took millions of years. ☐

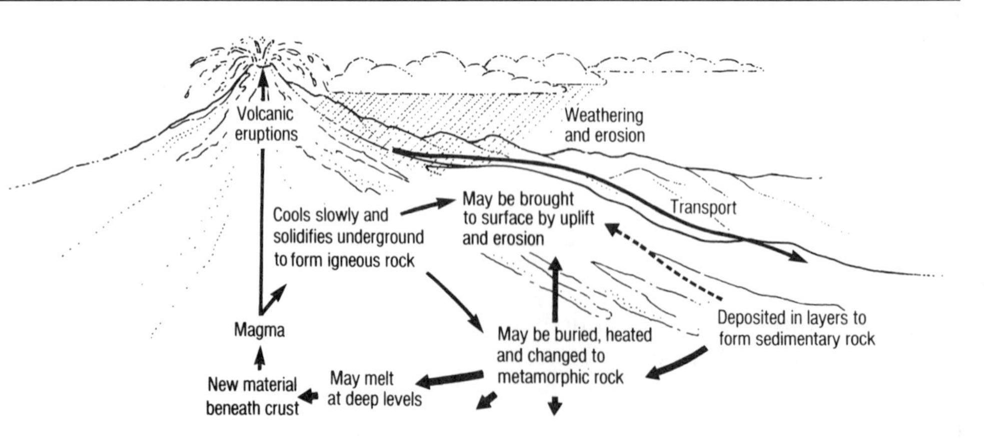

Draw a line to match the type of rock to the way it was formed.

Igneous	Deposited in layers (often from weathering processes)
Metamorphic	Liquid rock cools and solidifies and forms crystals
Sedimentary	Buried and changed by heat and/or pressure

B8 Features of the Earth, like mountains, caves and valleys, form over many millions of years.
- The movement of plates in the Earth's crust causes large-scale continental changes.
- The movement of water and ice over the surface causes more localised changes. ☐

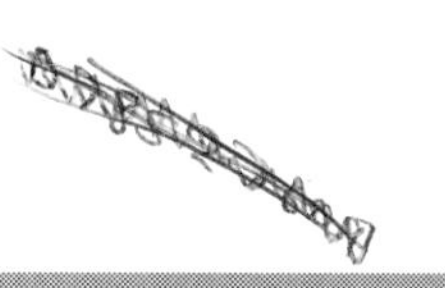

B9 In a solution, the ions become free to move. During **electrolysis**, ions are attracted to the oppositely charged electrode. Metal ions therefore move to the negative electrode (the cathode).

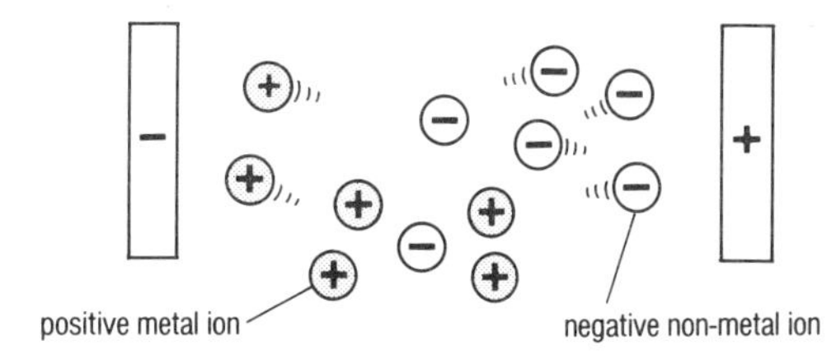

Electrolysis of copper chloride ❑

Put these phrases in order (1–4) to describe what happens to the non-metal ions during electrolysis.

Element formed .... Substance dissolves .... Attracted to anode .... Ions free to move ....

C1 Reactive metals give out a lot of energy when they react. Less reactive metals give out less energy. ❑

Complete the table, using the *water* column to guide you. Also try to work out why the strange sentence is written down the left-hand side.

	Metal symbol	Name	Reaction with water	Reaction with oxygen	Reaction with acid
Please	K		↑		
Send	Na				
Lord	Li				
Charles	Ca				
McLean	Mg				
A	Al				
Zebra	Zn		More energy released		
If	Fe				
The	Sn				
Lame	Pb				
Horse	H				
Cannot	Cu				
Munch	Hg				
Sweet	Ag				
Grass	Au				

C2 More reactive metals can push out less reactive metal ions from solution. This is called **displacement**. ❑

Underline or highlight key phrases.

A1 A material is useful because of its properties. ❑

Match the properties to the objects and add one other important property to each drawing. If you do not know the meaning of a property, use a dictionary.

high tensile strength high compressive strength high porosity high thermal conductivity high thermal stability low cost

metal	ceramic	plastic	metal	ceramic	nylon fibre
high compressive strength	high thermal stability	low cost	high thermal conductivity	high porosity	high tensile strength

A2 The periodic table arranges the elements to take account of their chemistry. Therefore elements in the same group have similar (but not identical) chemical properties. ❑

Complete the table for the other members of each group.

Group 1 (alkali metals)	Group 7 (halogens)	Group 8 (noble gases)
Li shiny metal, reactive	F non-metal gas, reactive	Ne non-metal gas, unreactive
Na	Cl	Ar
K	Br	Kr

A3 Atoms join together to make molecules. Molecules of an element contain the same type of atom; molecules of a compound contain different types. ❑

Where possible match the labels to the correct drawing.

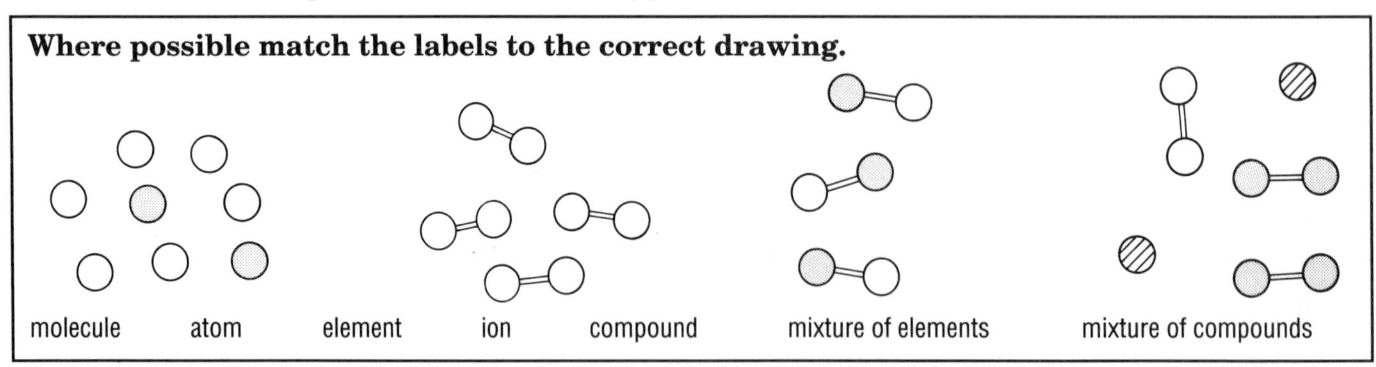

A4 Diagrams of particles can be used to illustrate what happens when a substance changes.

solid sodium chloride

The ions are joined in a stable lattice.

water molecules

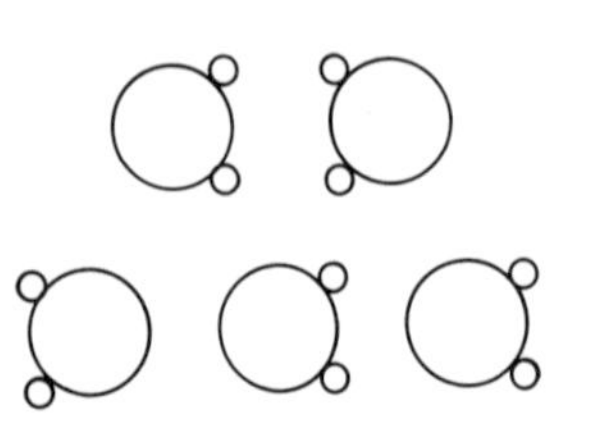

Add some water molecules.

sodium chloride solution

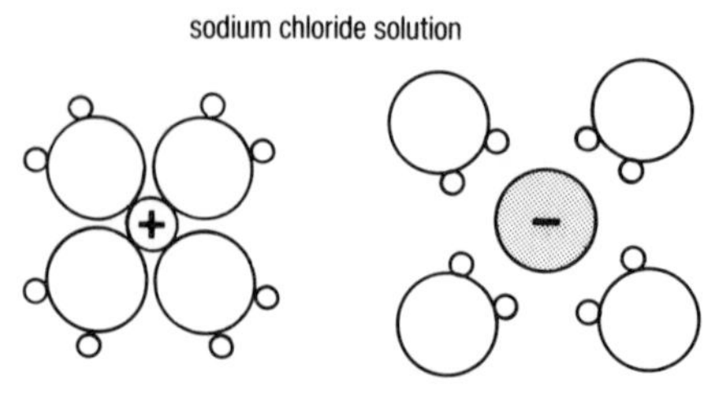

The ions become separated and surrounded by water molecules and are now free to move.

❑

A5 Atoms are made of electrons, protons and neutrons. ❑

Label the drawing below to show where these subatomic particles are found in an atom.

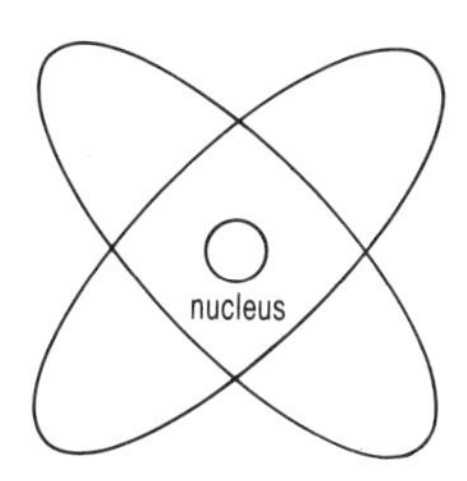

A6 All atoms of the same element contain the same number of protons. This number is known as the **atomic number**. ❑

Complete the following sentences.

Li contains 3 protons.

C contains . . . . protons.

Na contains . . . . protons.

Si contains . . . . protons.

A7 Elements and compounds can be represented by formulae because they are always made up in the same way. ❑

Complete the following table and add two more examples of your own.

Name	Formula	Is it an element or a compound?
water		
	CO_2	
	Na	
	O_2	
sodium chloride		

A8 Particles have kinetic (movement) energy. The particles from a perfume can move through spaces between the particles in the air. They **diffuse** and the smell reaches your nose. Particles of sugar diffuse through hot tea. ❑

Write down two other examples of diffusion of one type of particle through another.

A9 More heat energy will increase the spaces between particles, and eventually the substance will change state. ❑

Look back at pages 32 and 37 and then rewrite the previous sentence in your own words.

A10 more energy = faster movement = more space
higher temperature = higher pressure = larger volume

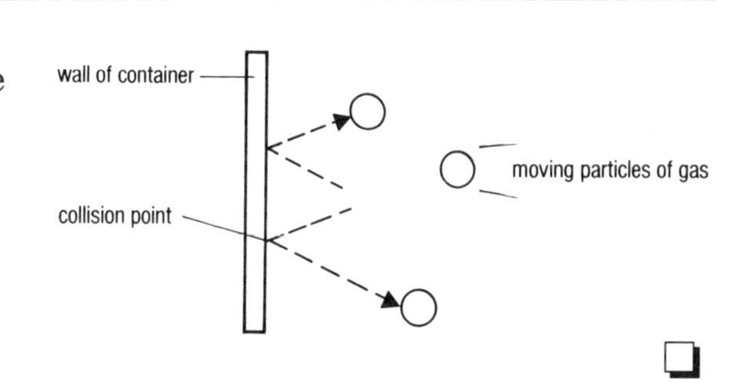

The gas particles are at high **temperature**. They have a lot of energy and so move fast, colliding with, and exerting **pressure** on, the walls of the container. They also take up a lot of space (**volume**). ❑

The connection between the three variables, temperature, pressure and volume, is shown by the formula:

$$\frac{\text{Pressure (atmospheres)} \times \text{volume (litres)}}{\text{temperature (Kelvin)}} = \textbf{constant value}$$

With no change in the volume, when the pressure goes up, the temperature **must** go up. If the pressure decreases, the temperature decreases too. ❑

Suppose we have 3 litres of gas at a pressure of 1 atmosphere and a temperature of 290 K. What will its volume be if the temperature rises by 20 degrees and the pressure remains the same? Follow this example.

$$\frac{P_1 \times V_1}{T_1} = \frac{P_2 \times V_2}{T_2}$$

$$\frac{1 \times 3}{290} = \frac{1 \times x}{310}$$

$$x = \frac{3}{290} \times 310$$

$$= 3{\cdot}2 \text{ litres}$$

Now, in the same way, work out the new volume of this gas if the pressure changes to 2 atmospheres and the temperature remains at 290 K.

B1 The rate of a chemical reaction varies with time. The graph, for example, shows the change in mass of calcium carbonate as it bubbles in hydrochloric acid. ❑

Mark on the graph:

the **start** of the reaction
the **end** of the reaction
the point with the **fastest rate** of reaction

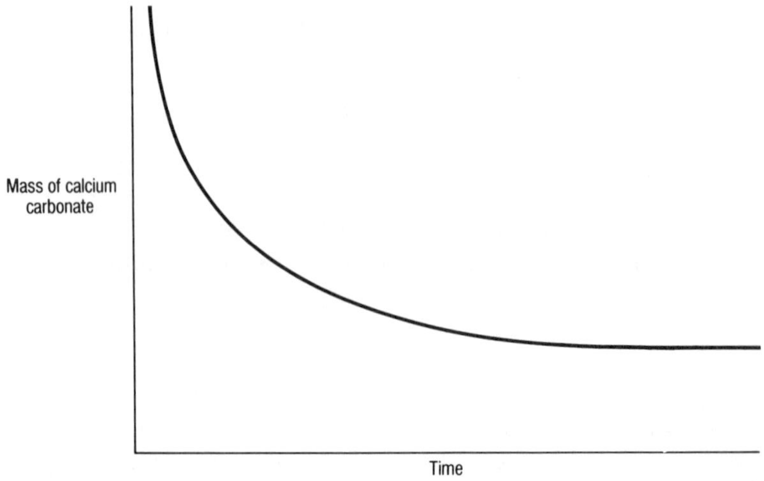

Mass of calcium carbonate

Time

B2 Particles must collide before they react. The rate of a chemical reaction can therefore be changed by changing the number of successful collisions. The rate of a reaction is increased by increasing the:

- concentration

crowded

More particles so more collisions

- surface area

More particles on outside so more collisions

- temperature

fast-moving particles

More energy to activate the collisions

❑

B3 Everyday life is affected by chemical ideas. For example, food cooks more quickly if it is divided into small pieces. ❑

Give one *everyday* example of increasing the rate of a reaction by increasing each of the variables below.

Concentration ....

Temperature ....

Surface area ....

Catalyst presence ....

B4 When rocks are broken down into smaller pieces it is known as weathering. Weathering can be caused by physical or chemical changes in the rock. ❑

Eight sentences are given below. Put them in the correct order so that four sentences give an example of chemical weathering and four sentences give an example of physical weathering. First label the sentences in the spaces on the right. Use P1–P4 for the examples of physical weathering and C1–C4 for chemical weathering. Then write out the sentences in order in the space provided below.

A After the rock has expanded and contracted several times, pieces of rock break off. ....

B These dissolved gases make the rain acidic. ....

C Rainwater enters cracks in the rocks. ....

D Acid rain can dissolve rocks such as limestone quite easily. ....

E Small pieces of rock are carried away by wind and rain. ....

F The dissolved limestone is carried away in the rainwater. ....

G When the water freezes, it expands and makes the cracks in the rock larger. ....

H Rainwater contains dissolved gases such as carbon dioxide, sulphur dioxide and nitrogen dioxide. ....

....

....

....

....

C1 Our knowledge of chemical ideas can also be used to manufacture new materials. For example, the more reactive a metal is, the more energy is needed to break up its compounds. Iron is extracted from iron ore by using carbon and heat. Aluminium, however, cannot be extracted from its ore in the same way. Instead, the ore must be melted and then electricity passed through it. The more reactive metal needs more energy to pull it away from the compound.
This partly explains why a metal like iron was known and used long before a metal like aluminium. ❑

Number these metals 1–7 in the order that they might have been discovered.

Na ...... Mg ...... Al ...... Hg ...... Au ...... Fe ...... Cu ......

C2 The pH scale is used as a measure of the acidity of a solution. ❑

Add the following to the pH scale in the correct position.

strong acid	weak acid	neutral	hydrochloric acid	strong alkali
water	sodium hydroxide	lemon juice	bleach	weak alkali

1	2	3	4	5	6	7	8	9	10	11	12	13	14

C3 Some substances react with acids and neutralise them. Neutralisation reactions follow a pattern.

Acid + Neutraliser ⟶ a Salt + water or hydrogen ❑

From the list of substances make up two word equations that describe a neutralisation reaction.

sodium hydroxide	magnesium	magnesium chloride	magnesium nitrate	hydrogen
water	sulphuric acid	sodium sulphate	sodium chloride	nitric acid

1 .... + ... ⟶ .... + ...

2 .... + ... ⟶ .... + ...

C4 Neutralisation reactions are sometimes helpful. ❑

Sting

Acid soil

lime

Sketch another example of your own.

ATTAINMENT TARGET 4

Physical Processes

Big ideas

A Electricity and magnetism can be controlled to serve human needs.
B The effects of forces can be used to solve problems in our everyday lives.
C Sound and light can be detected by humans and have many properties that we can make use of.
D Day length, the seasons and many other things are the result of the Earth's position in the solar system.
E The energy we depend on from day to day comes from many sources, some of which are running out.

Key ideas
Write 2 or 3 key words from each paragraph in this column

A1 Materials that conduct electricity are called **conductors**.
Materials that do not conduct electricity are called **insulators**. ❑

Complete the table by writing *conductors* or *insulators* in the correct place.

insulator ✓	conductor ✓
wood, plastic, glass	iron, copper, aluminium

A2 The path that electricity takes is called a **circuit**. ❑

A3 Symbols are used to show the parts of an electric circuit. ❑

Draw a line to join the name of each part of a circuit to its symbol.

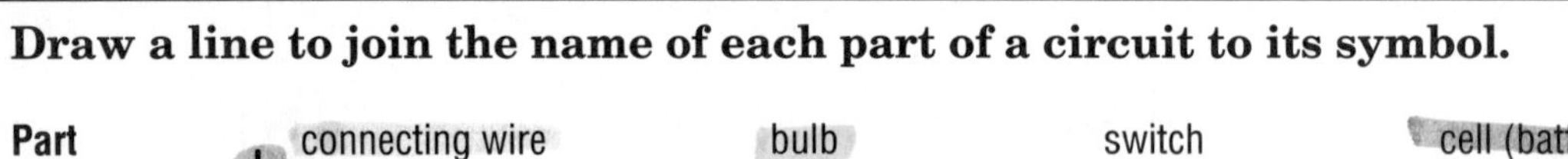

Part connecting wire · bulb · switch · cell (battery)

Symbol

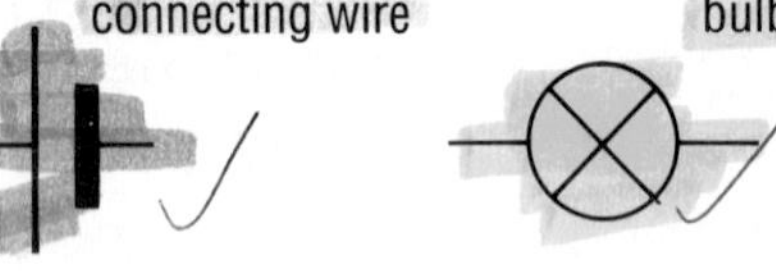

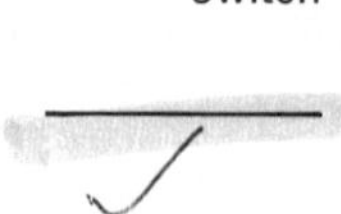

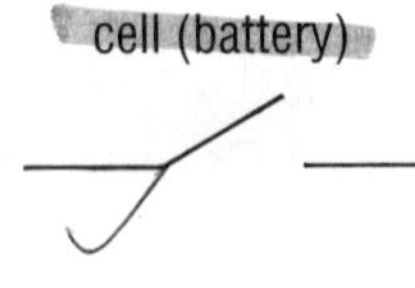

A4 A complete circuit is needed for a bulb to light or a bell to ring. ❑

Tick the circuits where the bulb would light.

A. ✓

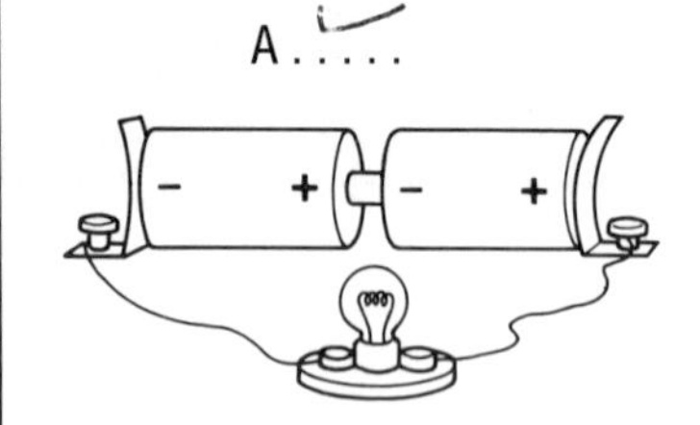

B.....

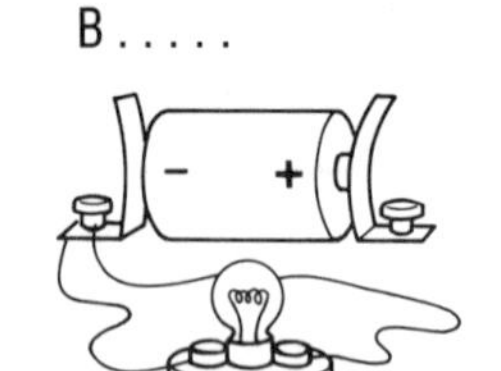

C.....

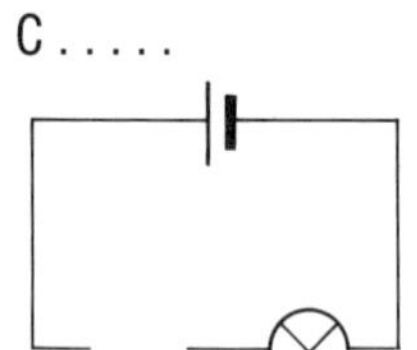

D. ✓

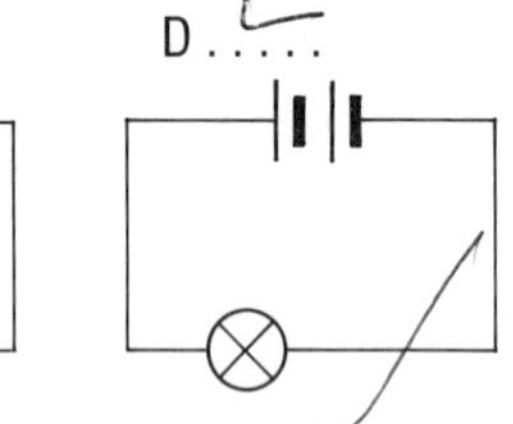

B1 **Forces** can affect the position, speed of movement and shape of an object. For example, think about what happens to a soft ball when it is hit with a bat. ❑

Complete the diagrams with the words *direction, shape, speed.*

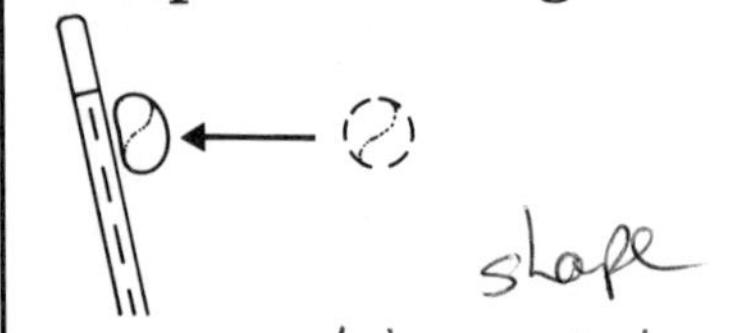

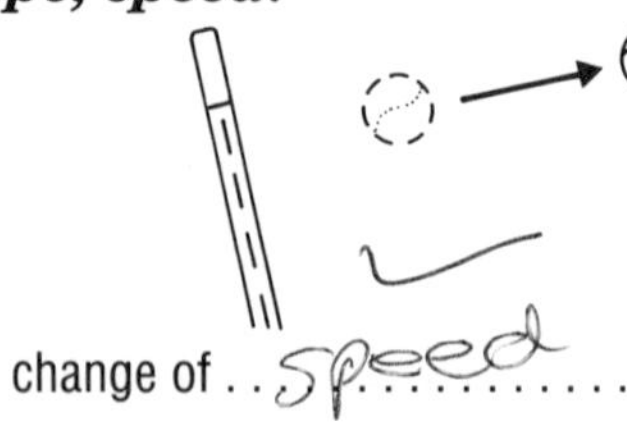

change of ..direction (shape) change of ..shape (Direction) change of ..speed ✓

C1 Sounds can bounce off (reflect off) hard surfaces. This can cause an echo. ❑

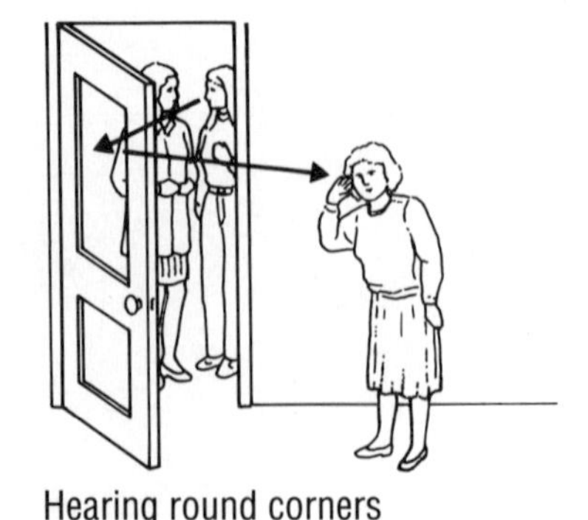

Hearing round corners

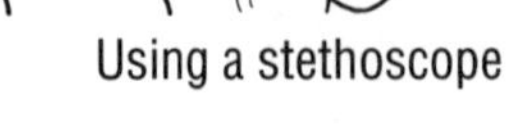

Using a stethoscope

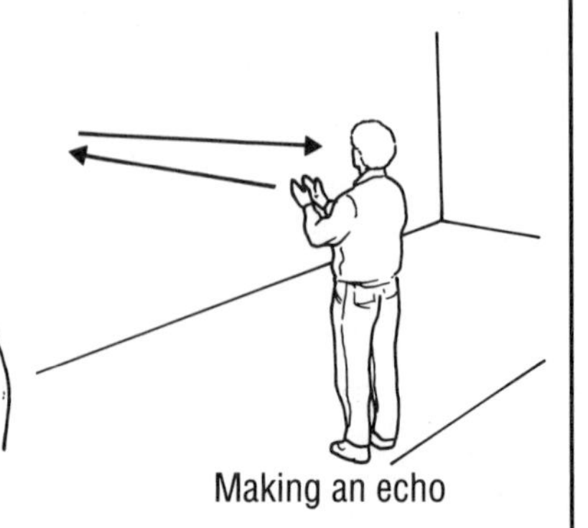

Making an echo

Key ideas
Write 2 or 3 key words from each paragraph in this column

C2 Light can be made to change direction by reflecting it off a mirror.

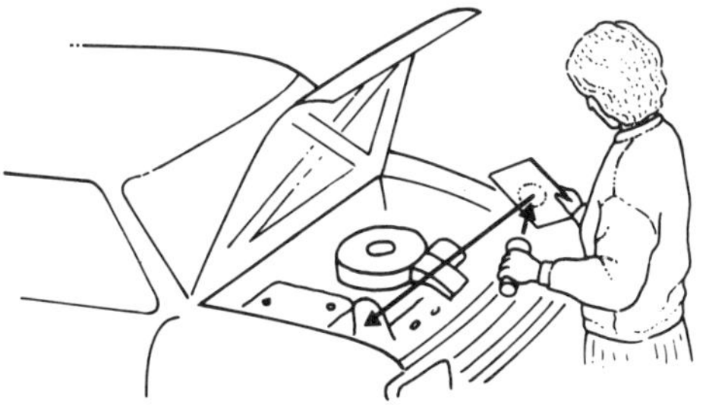

Looking into a dark corner

D1 The Earth is a planet in the solar system and it goes round a star we call the Sun. The Moon goes round the Earth.

Write *Earth, Moon, Sun* in the correct place.

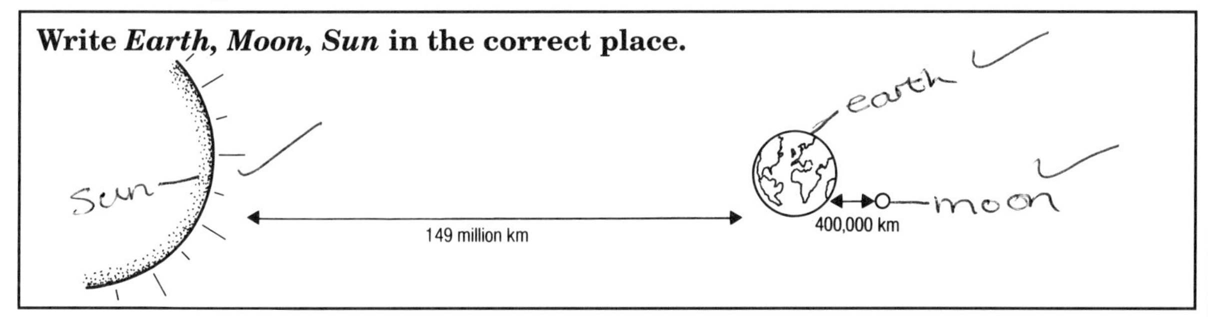

D2 From Earth, the Moon looks different at different times of the month. A pattern is repeated every 28 days.

Label the *full moon* and the *new moon.*

D3 From Earth, the position of the Sun in the sky changes throughout the year.

Write *winter, spring, summer* in the correct place.

E1 A **fuel** is a substance which burns, using oxygen, to release energy.

E2 A range of fuels are used at home for cooking and heating. During cooking, the heat energy released from the fuel is passed to the food.

Underline all the fuels that are used in houses for cooking or heating.

coal uranium wood gas petrol plastic oil

E3 Electrical energy can be used for cooking and heating. Electricity can be made from a range of fuels.

Key ideas
Write 2 or 3 key words from each paragraph in this column

A1 The flow of electricity through a conductor is called an electric **current**. ☐

A2 Circuits can have components in **series** (all on the same branch) or in **parallel** (on different branches). ☐

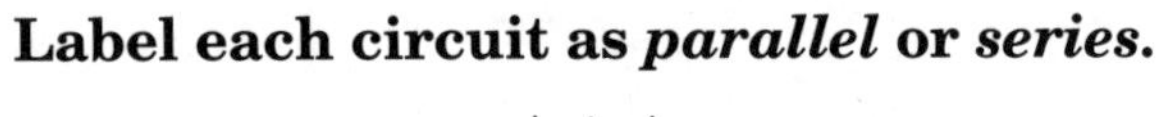

Label each circuit as *parallel* or *series*.

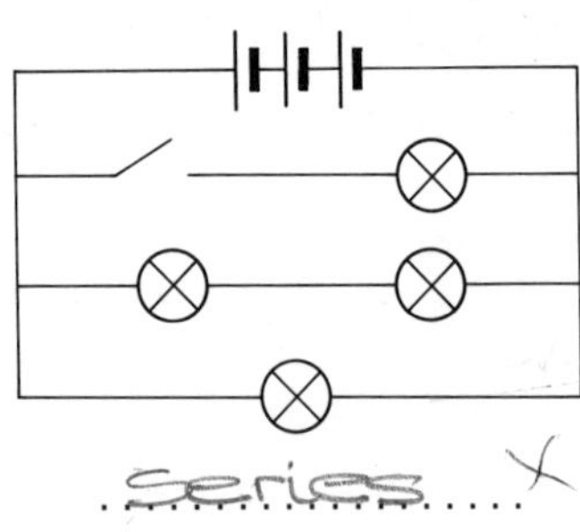

Series X

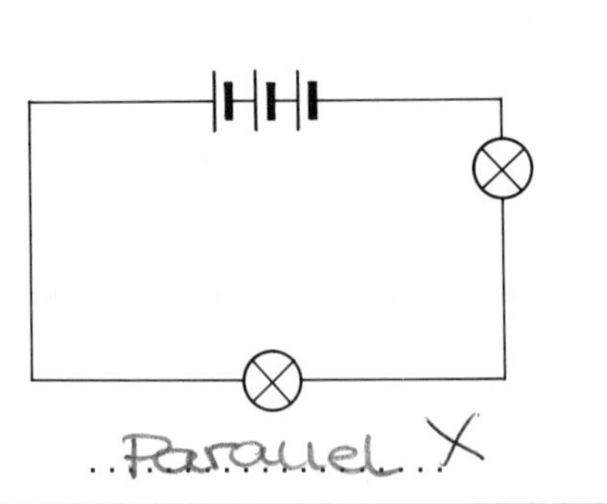

Parallel X

A3 Switches can be used to control what a circuit does. For example, in this circuit, the switch switches on the motor and the lights. ☐

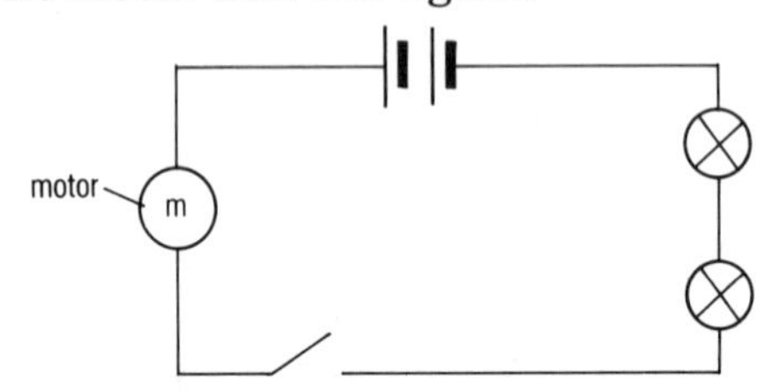

Which switch or switches would have to be on in this circuit to: (a) turn on the buzzer only? (b) turn on the lights and the buzzer?

(a) A

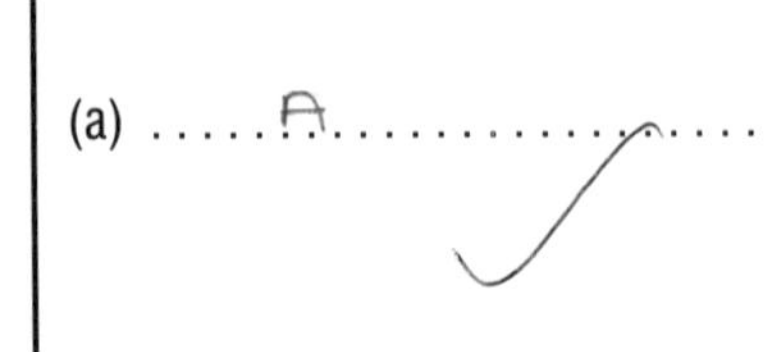

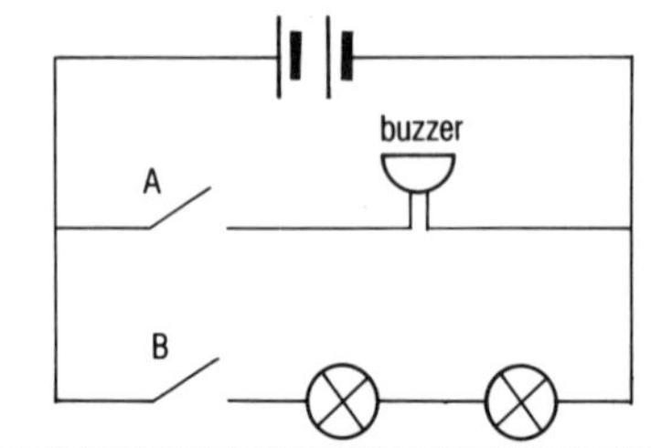

(b) A and B

B1 Force is measured in Newtons (N). ☐

B2 Simple machines seem to make work easier. For example: Less force is required to lift the crate with a pulley.

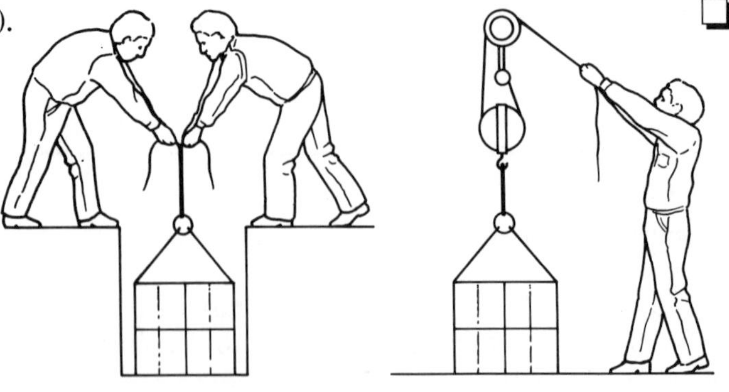

Less force is needed to lift the paving slab using a lever. ☐

Key ideas
Write 2 or 3 key words from each paragraph in this column

B3 More than one force can act on an object at the same time. Forces can also act in different directions: for example, when you have to push a car. ❑

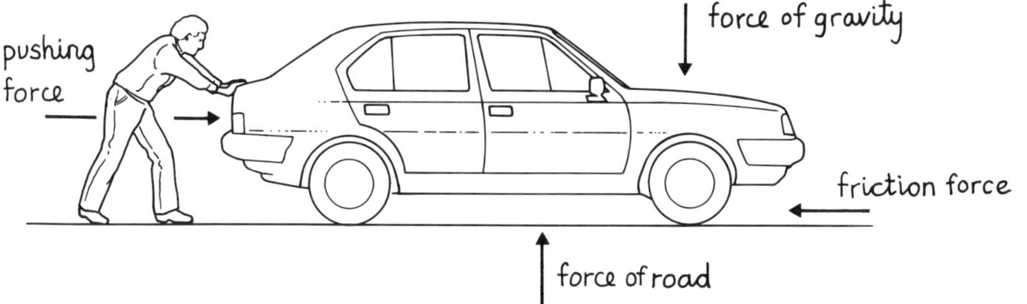

Draw arrows to show the directions of forces acting on the cannon ball.

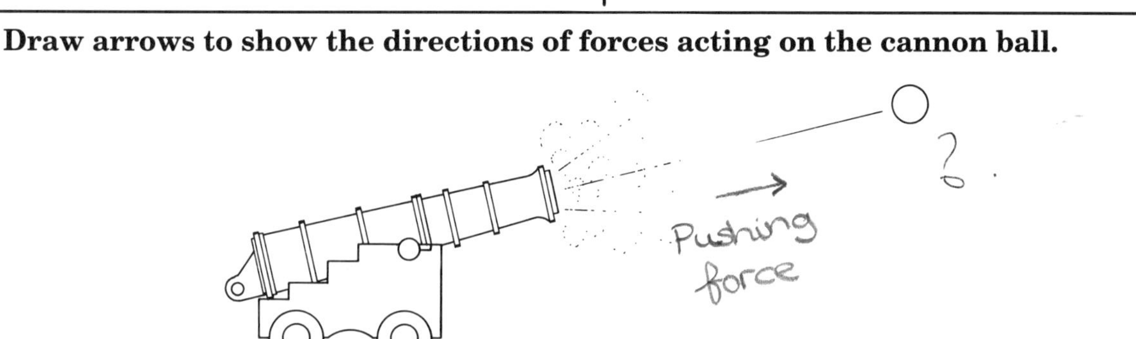

B4 The forces acting on an object can be in balance. For example:
When a ball is at rest. When a ball is floating. When a ball is moving at a steady speed. ❑

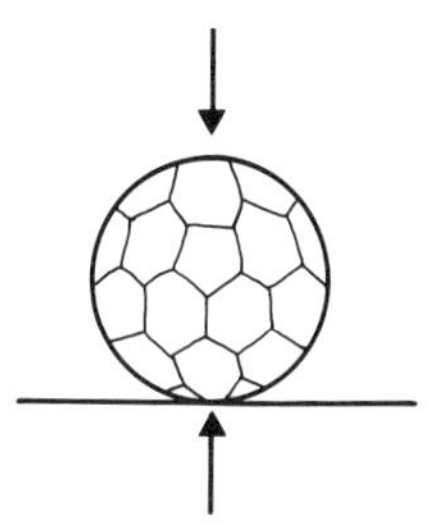

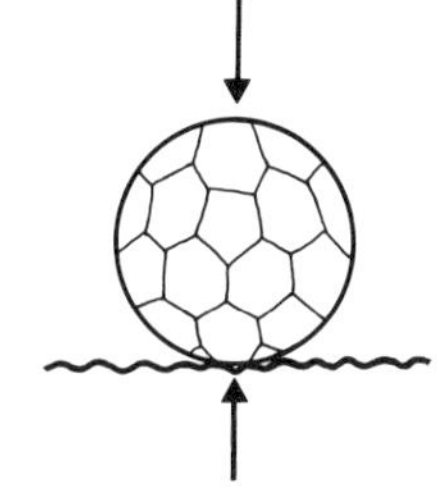

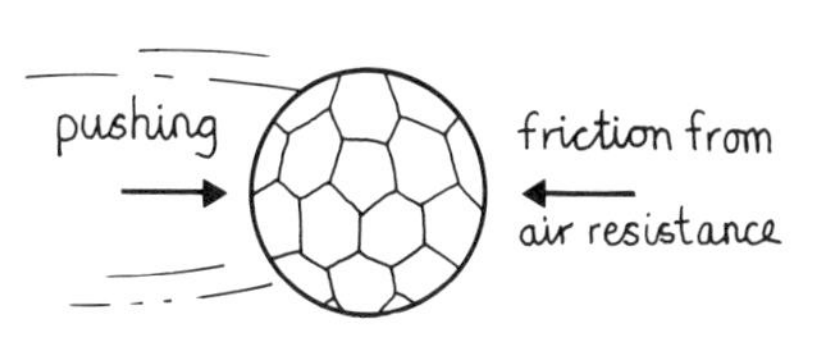

C1 Light travels much faster than sound. ❑

Complete the sentences with the words *before* or *after*.

Lightning is seen . . before . . . . thunder is heard.

The sound of a bat striking a ball is heard . . . after . . . . the ball is seen to move.

C2 Sounds can travel through different materials such as air, water and metal. Sounds cannot travel through a vacuum. ❑

Key ideas
Write 2 or 3 key words from each paragraph in this column

D1 As planet Earth revolves around the Sun, it spins or rotates. This causes the pattern of day and night. ❑

D2 It takes the Earth one year to make one complete revolution of the Sun. ❑

D3 The Earth spins at an angle on its axis. This causes the differences in day length which we notice in Britain at different times of the year. ❑

Complete the diagram with the words *summer, winter, short days, long days* to show what people in Britain would experience.

North
Britain
Sun
North
Britain

..Summer.......... ✓ ...winter........... ✓

..Long...days..... Short.days.......

D4 The length of your shadow depends on the position of the Sun in the sky. ❑

Complete the following sentence, using the words *long*, *short*.

On a sunny day I have a .Short...... shadow in morning and a .Long....... shadow at lunchtime.

E1 There are several forms of energy: for example, heat energy, sound energy, light energy, stored energy, movement energy and electrical energy. ❑

E2 Energy can be transferred from one energy form to another. An **energy transfer** is needed to make things work. ❑

Complete the table to show the energy transfers involved.

clockwork toy	stored to ...Clock.work...toy..................
battery-powered car	.~~Steam~~.engine to .electric.motor. to movement
steam engine	heat to ..steam....engine....................
electric motor	battery-powered.car to ..electric.motor...

E3 Energy is measured in Joules (J). 1000 J = 1 kJ (kilojoule). ❑

Key ideas
Write 2 or 3 key words from each paragraph in this column

A1 An insulating material can be given an electric charge by rubbing. This is called **static electricity**. For example, when you rub a comb it becomes charged and can attract bits of paper. ❑

A2 In any electric circuit there must be:
- an electric charge
- a conductor – the wires
- a force to move the charge – supplied by the battery or cell. ❑

A3 Some electrical components can be used to control the flow of charge in a circuit. ❑

Draw a line to match each component with what it does in a circuit.

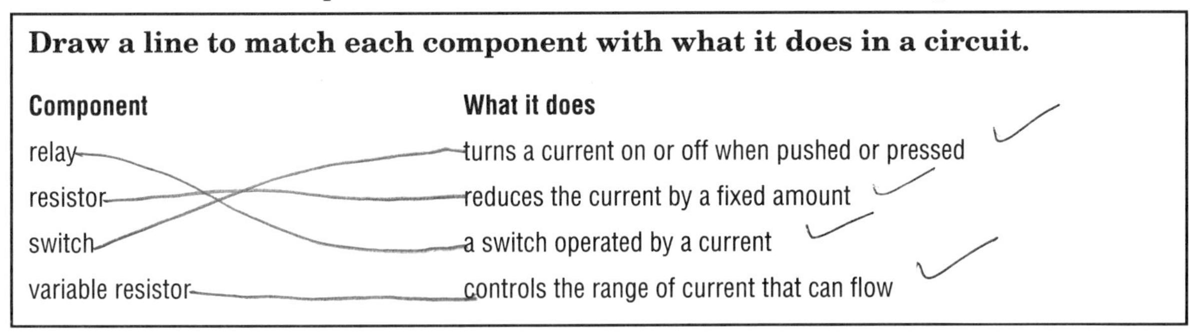

A4 Current carries electrical charge round a circuit. Current is not used up by components in a circuit. ❑

B1 The movement of an object depends on the size and direction of the forces acting on it. For example, when you push a car:
- just starting to move – overcoming the frictional force
- moving more quickly (accelerating) – pushing force is greater than friction
- moving at a steady speed – pushing force equals frictional force ❑

Complete diagrams 2 and 3 to show the forces acting on a parachutist at different stages in a jump.

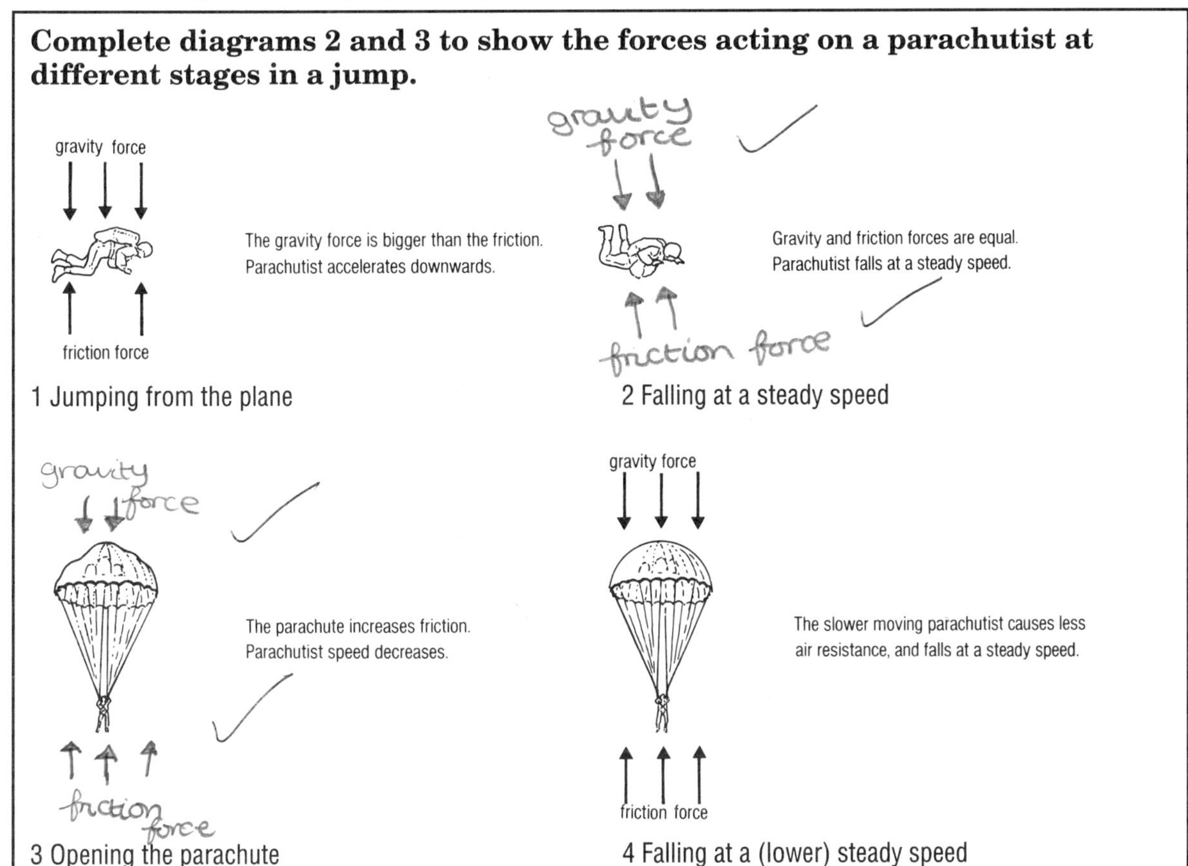

Key ideas
Write 2 or 3 key words from each paragraph in this column

C1 We see objects because some light is scattered from them and reaches our eyes. ❑

Draw three arrows to show the path of reflected light to the eye in picture 1, and two arrows to show the same thing in picture 2.

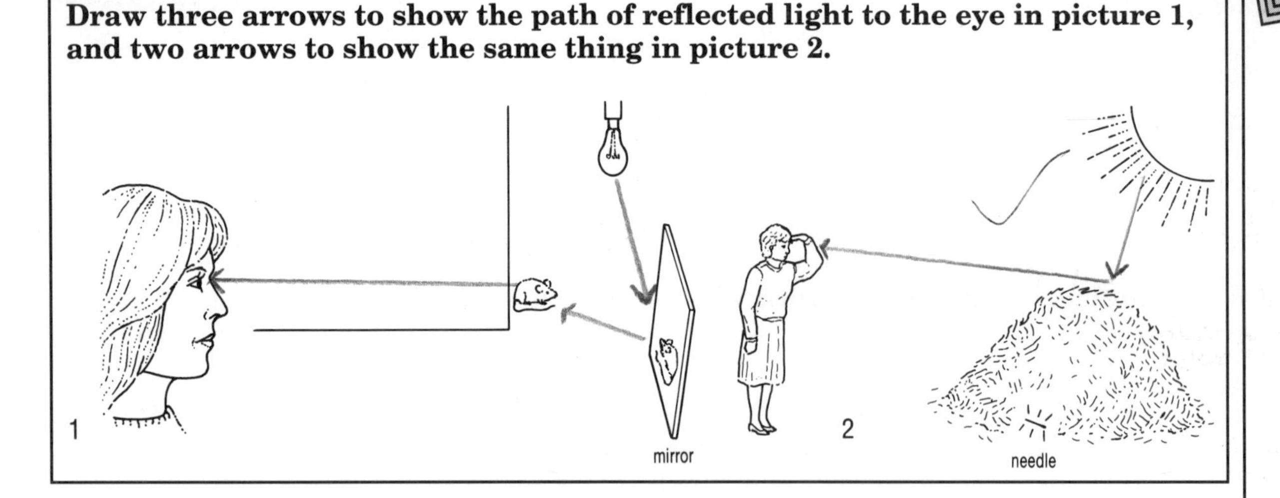

C2 Objects that produce sounds have vibrating parts. The vibrations are transferred to the air and pass through the air as **sound waves.** For example, we can hear sounds from a loudspeaker because changes in the electric current make parts of the loudspeaker move back and forth creating sound waves in the air. ❑

Put an X on the vibrating parts of the objects below.

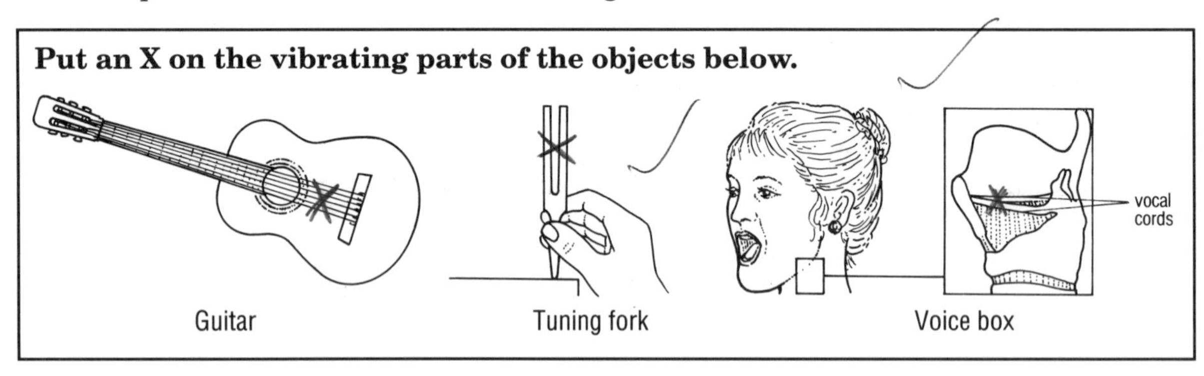

Guitar Tuning fork Voice box

C3 Rulers and milk bottles can be used as simple musical instruments as shown in the pictures below. ❑

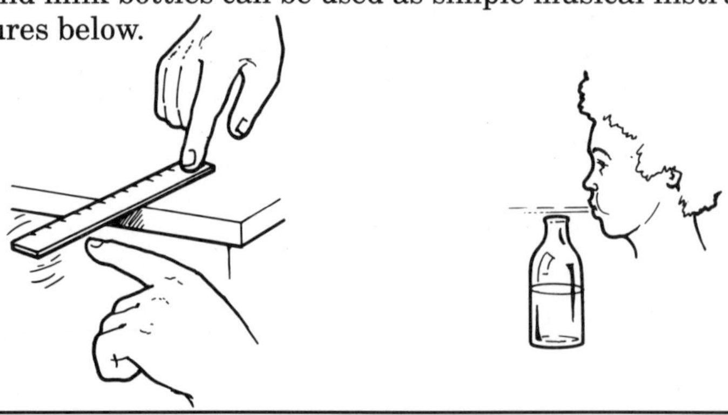

Write a sentence to explain how you would make (a) a louder sound and (b) a higher sound with each instrument.

Ruler

(a) louder by plucking it harder

(b) higher pitch by leaving less off the table

Milk bottle

(a) louder by blowing harder

(b) higher pitch by closing the hole more

Key ideas
Write 2 or 3 key words from each paragraph in this column

D1 The solar system is made up of the Sun and nine orbiting planets. These are, from the Sun outwards: Mercury, Venus, Earth, Mars, Jupiter, Saturn, Uranus, Neptune and Pluto. ❑

Label the planets in the diagram. Below each planet's name write the time taken for one orbit of the Sun. Choose from: 248 years, 165 years, 84 years, 30 years, 12 years, 687 days, 365 days, 225 days, 88 days.

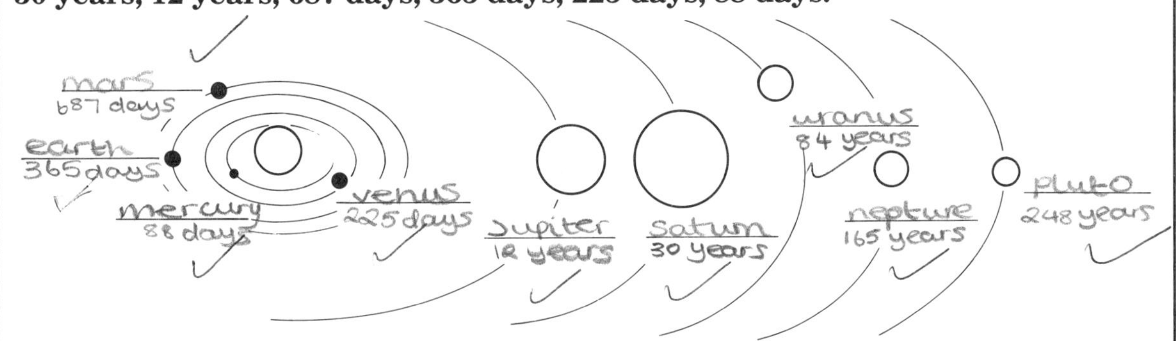

D2 Human exploration of space is limited. Unmanned spacecraft have left the solar system, but humans have not landed on any of the other planets. In 1969 humans landed on the Moon, which is not a planet. ❑

D3 The Earth spins or rotates from west to east as it orbits the Sun. It is also tilted at an angle as it spins. This explains why we have seasons and night and day. ❑

Look at the four diagrams of the Earth below. In each case decide whether it is summer or winter and day or night in Britain.

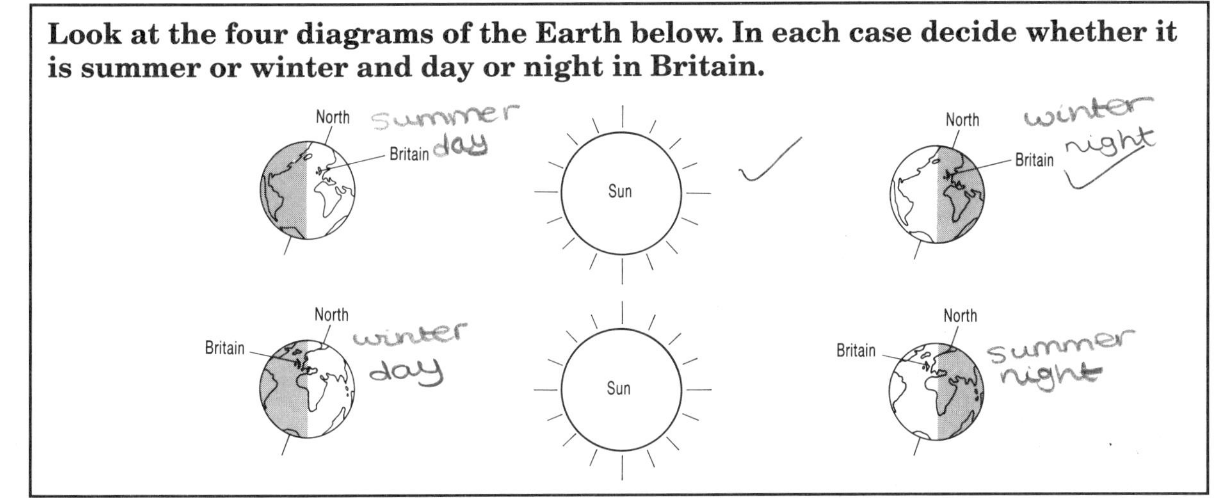

D4 The Sun is a star. Like other stars, it makes light and heat energy. The Moon and the planets are visible to us because sunlight is reflected from their surfaces. ❑

Draw two arrows to show the path of reflected light from the Sun to your eye.

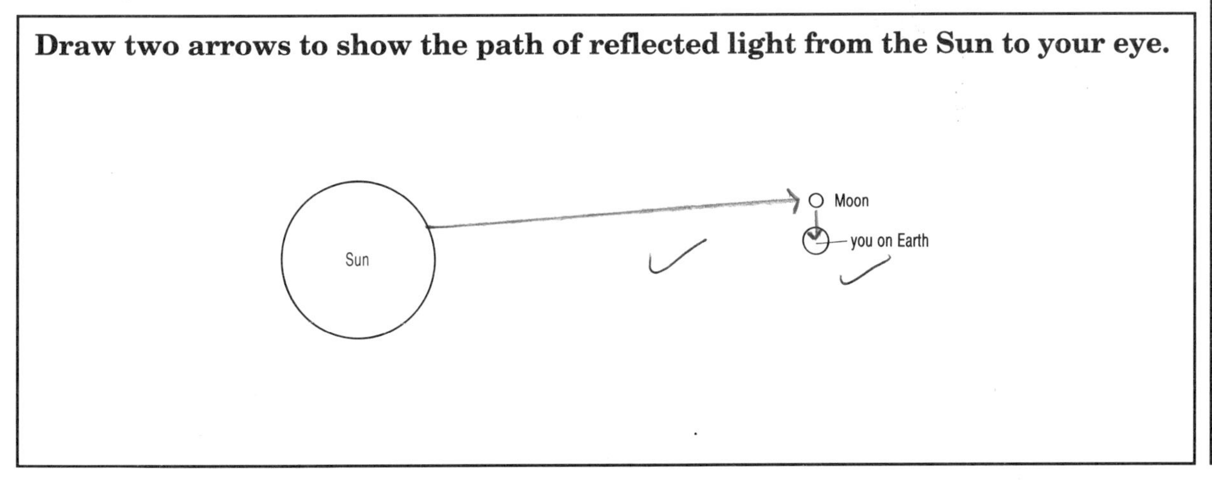

Key ideas
Write 2 or 3 key words from each paragraph in this column

E1 In any process, an energy transfer takes place. For example, when water is boiled in an electric kettle, electrical energy is transferred to heat energy. ❑

Complete these energy transfers.

ringing a door bell .... to ..sound..........

pedalling a bike .... to ..movement......

using a solar cell to drive a motor ..heat.... to .... to ..movement..

E2 Electricity can be generated from renewable and non-renewable energy sources. ❑

Write *renewable* or *non-renewable* beside each set of examples.

coal, gas, uranium, oil .non-renewable.......

moving water, wind, sunshine ..renewable......

E3 Non-renewable energy sources must not be wasted because they cannot be replaced. In Britain and other parts of the world, electricity is produced from:

- hydroelectric power
- coal
- oil
- gas
- nuclear fuel

❑

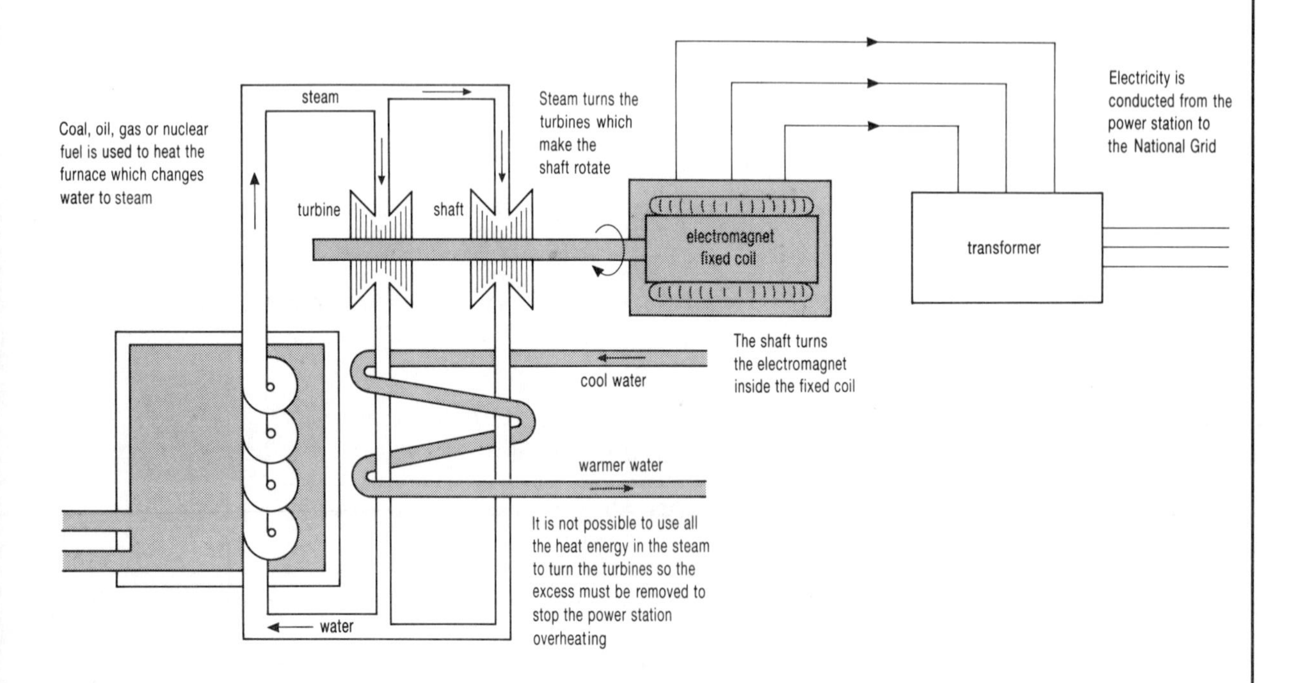

Underline or highlight key phrases.

A1 Different materials allow electric charge to flow through them to different extents. This is called **resistance**. ❑

Complete the sentences with the words *high* or *low*.

Copper is used in electrical wiring because it has a . . . . resistance.

A plastic sleeve is wrapped round the copper wire because the plastic has . . . . resistance.

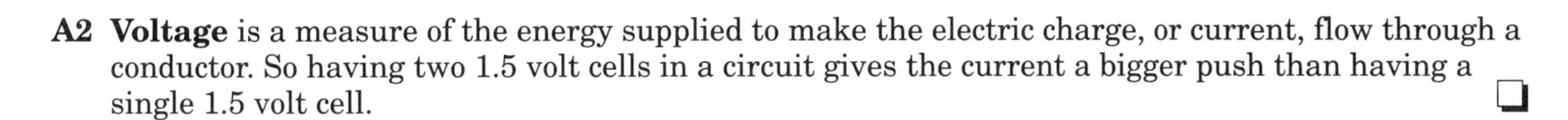

A2 **Voltage** is a measure of the energy supplied to make the electric charge, or current, flow through a conductor. So having two 1.5 volt cells in a circuit gives the current a bigger push than having a single 1.5 volt cell. ❑

A3 The higher the resistance of a circuit, the lower the current will be. The brightness of a bulb can be used as a measure of the current flowing in a circuit. ❑

Tick the circuit where the current is higher.

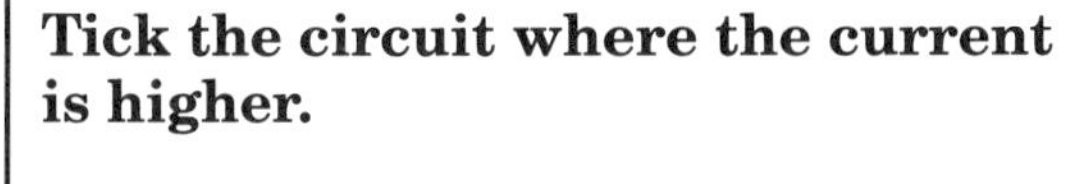

A . . . . .

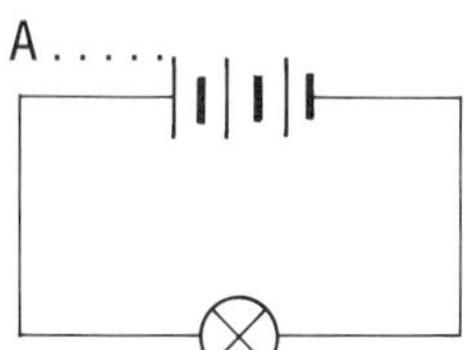

B . . . . .

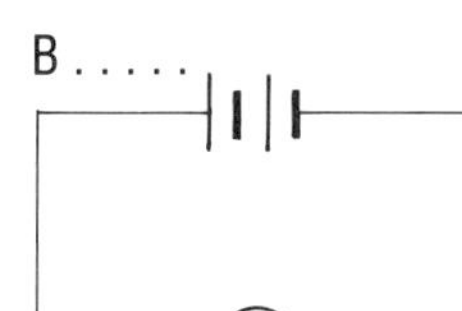

Tick the circuit where the current is higher.

A . . . . .

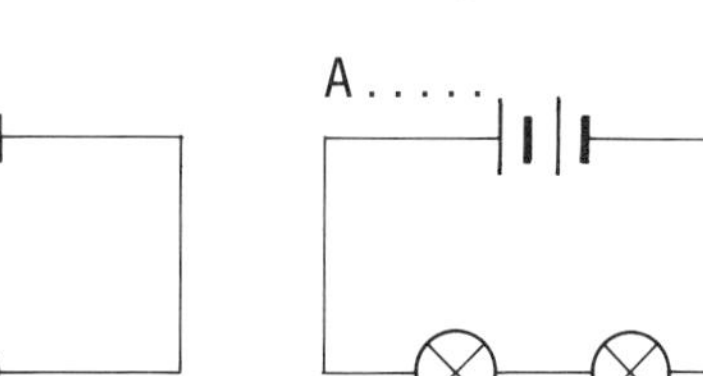

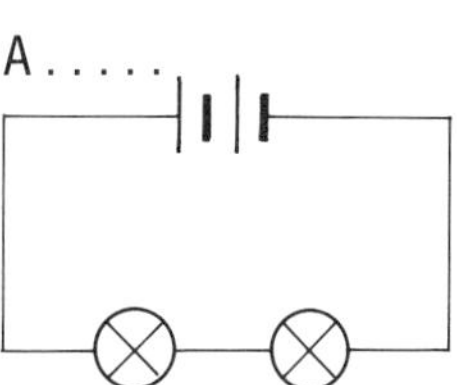

B . . . . .

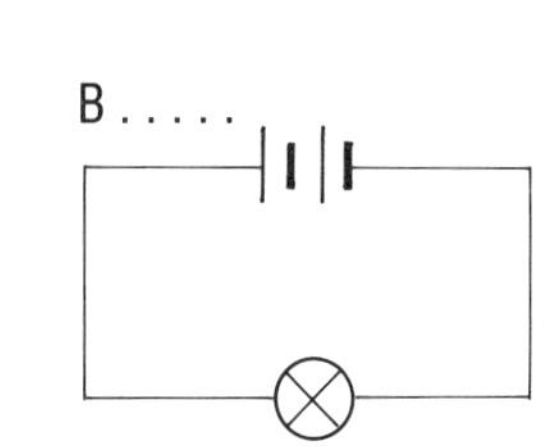

A4 In a series circuit, the current, measured using an ammeter, is the same at all points in the circuit. In a parallel circuit, the current divides into each branch. The size of current in each branch depends on the resistance in each branch. ❑

A5 When insulating materials are rubbed, they can have a positive or negative charge.
- Objects with the same charge repel.
- Objects with opposite charges attract. ❑

Draw arrows to show the expected direction of movement of the charged rods.

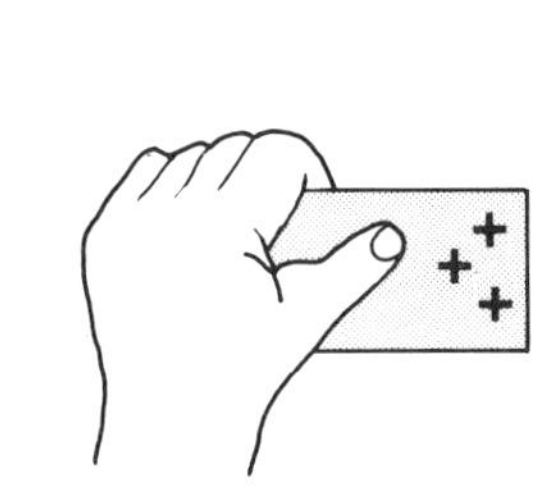

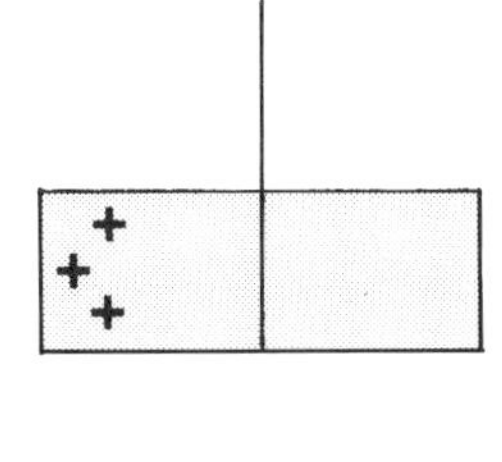

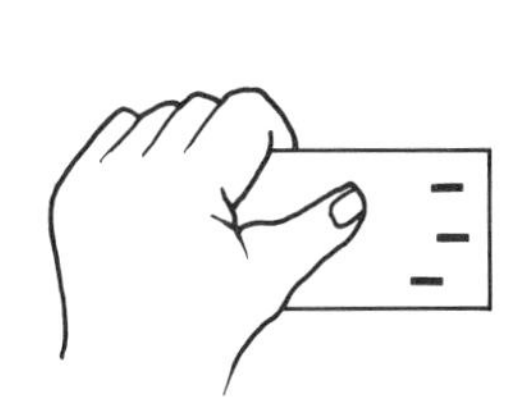

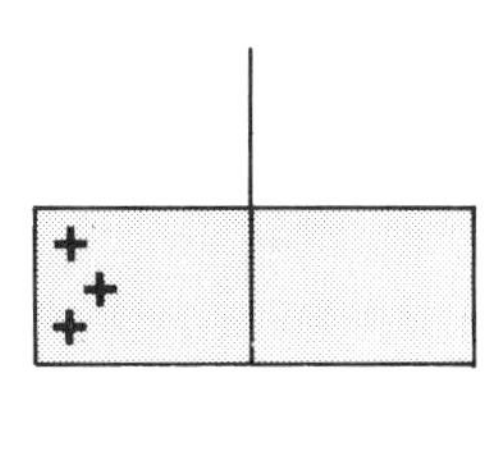

A6 An object will remain charged with static electricity until it touches a conductor. The static charge suddenly moves and is discharged. ❑

Write down two examples of static electricity being discharged.

B1 When a big force acts over a small area, we say it exerts a large pressure. For example, a flat-heeled shoe exerts less pressure on the ground than a stiletto-heeled shoe. ❑

In each pair circle the drawing where *less* pressure is being exerted.

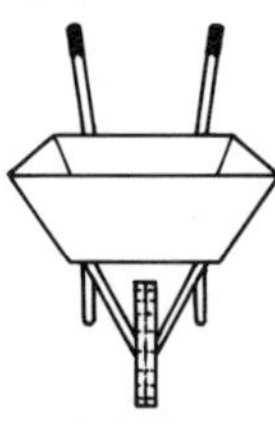
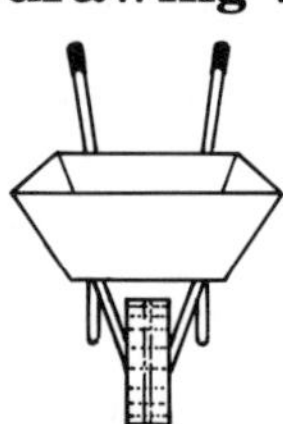
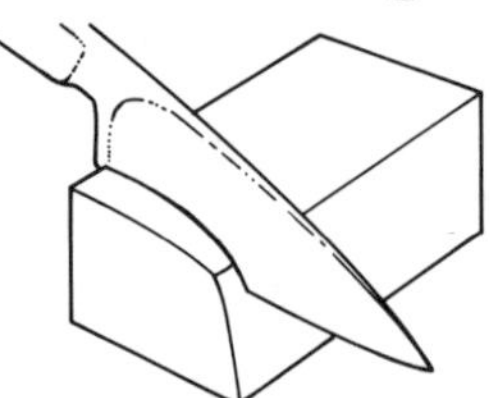
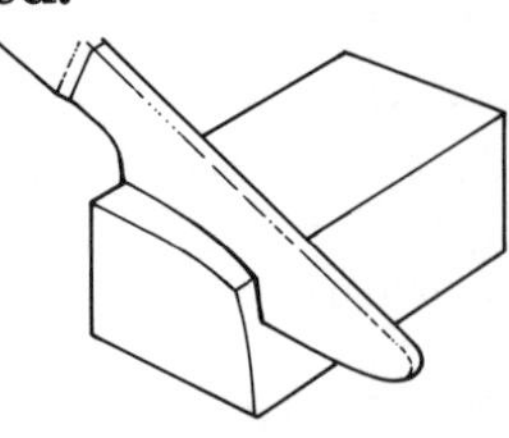

B2 Forces can make materials bend, stretch or twist. This idea is important in the design of bridges. ❑

B3 A car braking system depends on pressure. Foot pressure on the brake pedal is transmitted through brake fluid to the brake pads which are pushed against the brake drum or disc. ❑

B4 An object floats when the downward force balances the upward force. An object sinks when the downward force is greater than the upward force (upthrust). ❑

B5 The mean or average speed of a car can be calculated from measurements of distance travelled and time taken. ❑

Tick the correct formula and calculate the speed of the car.

speed = distance × time

speed = distance/time

speed = time/distance

A car travelled 200 km in 2½ hours.

mean speed =

B6 The distance required for a car to stop depends on many factors. ❑

For each factor, underline the condition that results in the *shorter* stopping distance.

Factor	Condition		Factor	Condition	
road surface	wet	dry	car tyres	deep tread	shallow tread
car speed	slow	fast	car brake pads	rough	smooth

C1 Light behaves like a wave. When light passes from one material to another at an angle, it changes direction. This is called **refraction**. Refraction occurs because light travels at different speeds in different materials. The denser the material the slower it travels and the more it is refracted. ❑

Complete these sentences explaining what happens to light.

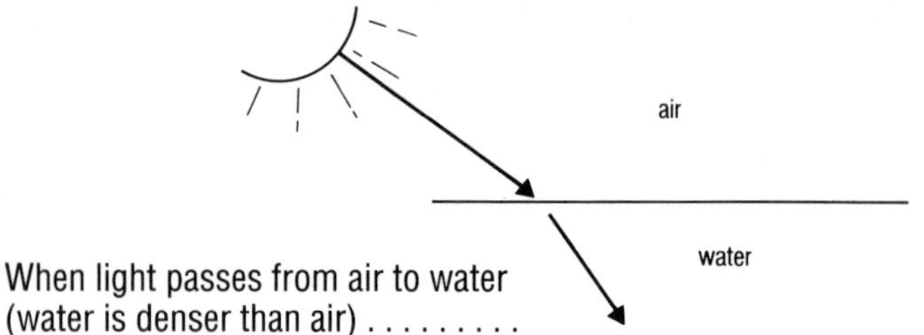

When light passes from air to water (water is denser than air) . . . .

. . . .

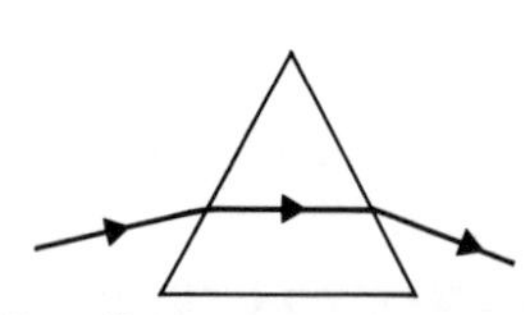

When light passes from air through a denser transparent prism . . . .

. . . .

Complete the diagram to show the path of light through a prism which is made of a denser material than the prism shown on page 56.

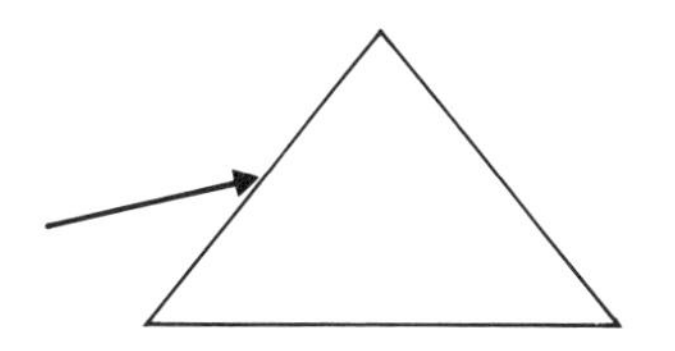

C2 We hear because sound waves enter the ear where they are changed to electrical impulses which are sent to the brain. ❑

Complete the table.

Name of part	Function of part of ear
	transmits sound to middle ear
	transmits and amplifies sound
	changes vibrations into electrical impulses
	takes electrical impulses to the brain
	keeps air pressure in outer and middle ear equal

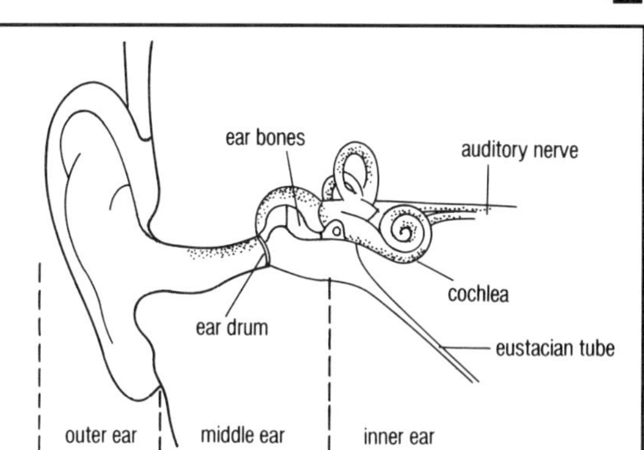

C3 Sounds can be described by how loud they are and by their pitch – whether the note is high or low. Sound can be 'seen' on an oscilloscope where it appears as waves.

Loudness depends on the energy of the sound wave. This is shown by the size or amplitude of the wave.

Pitch depends on how fast the vibrations are. The speed of vibration is called the frequency and it is measured in hertz (Hz). ❑

Write loud or soft under the correct diagram.

Write high or low under the correct diagram.

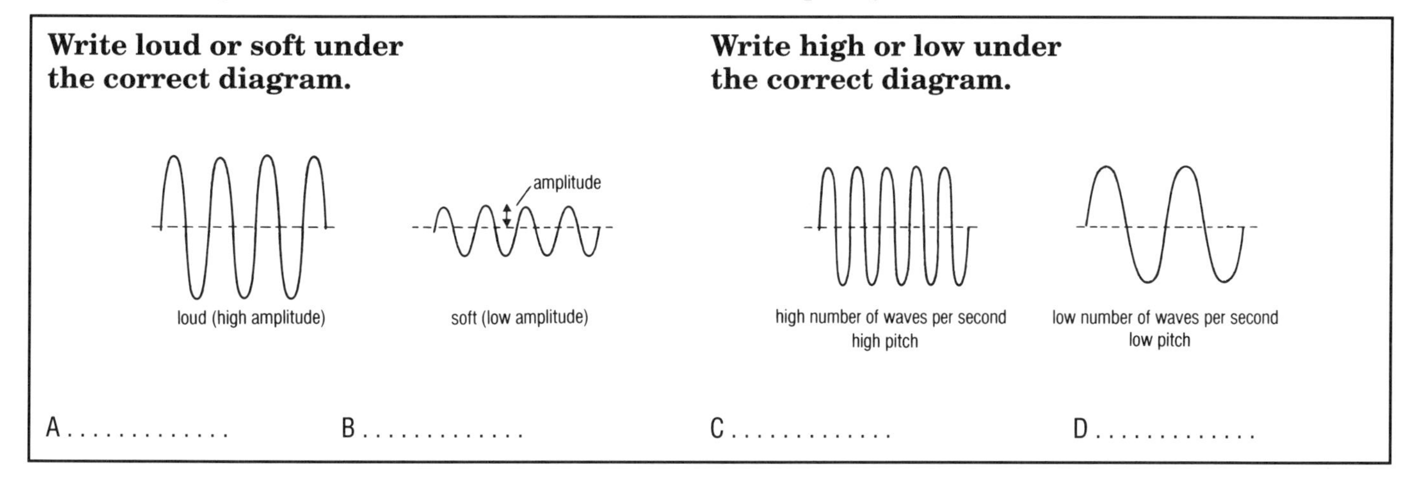

A............. B............. C............. D.............

C4 The range of human hearing is from about 20 hertz to about 20 000 hertz (20 kilohertz). ❑

D1 Stars in the night sky are not all equally bright. This could be due to their distance from the Earth, their age, size and temperature. ❑

D2 Satellites are sent into space to study conditions on Earth, like the weather, and to observe the Universe. ❑

D3 Our Sun is one of many millions of stars in a galaxy called the Milky Way. The Milky Way is one of millions of galaxies in the Universe. The Universe is made up of many clusters of galaxies which are separated by vast distances of empty space. There are thought to be nineteen galaxies in the local cluster containing the Milky Way. ❑

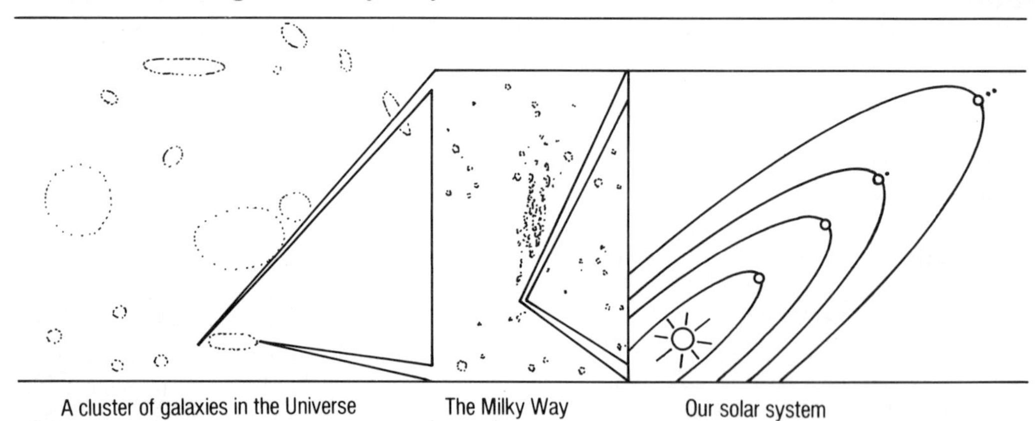

E1 Energy transfer is needed for people, machines and other devices to work and for change to take place. It happens all around you. ❑

Give one example of an object or device where there is an energy transfer by:

- light ....
- sound ....
- electricity ....

E2 In any energy transfer, some energy is changed into a form that is not useful. This energy is wasted, but it is not lost. This means that energy is conserved – the total energy put into a system is the same as the total energy which comes out of the system. ❑

Underline the wasted energy in the examples.

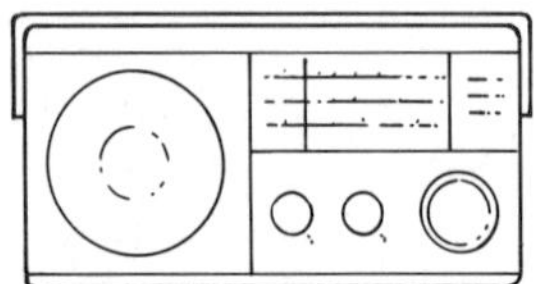

electrical energy → sound energy + heat energy

stored chemical energy → heat energy + light energy + sound energy

E3 Wasteful transfers of energy in machines can be reduced by reducing friction with, for example, oil or grease. ❑

E4 Most renewable and non-renewable energy sources can be traced back to the Sun. ❑

Complete the table with the correct energy source. Choose from:
fossil fuels* (coal, oil, gas), *wind power, solar power.

Energy source	Link with the Sun
	Solar cells trap sunlight and change solar energy into electrical energy.
	Made from dead plants and animals, once part of a food chain which started with solar energy.
	Air at two different places is heated unevenly by the Sun. As the warmer air rises, cooler air rushes in to replace it, causing wind.

Underline or highlight key phrases.

A1 Magnets have a north and a south pole. ● Opposite poles attract. ● Like poles repel. ❑

Complete the diagrams with the words *attract* and *repel*.

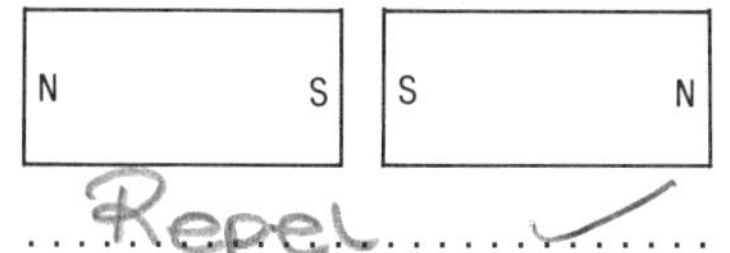

Repel ✓

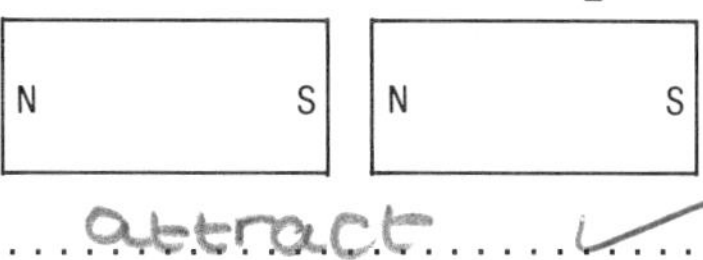

attract ✓

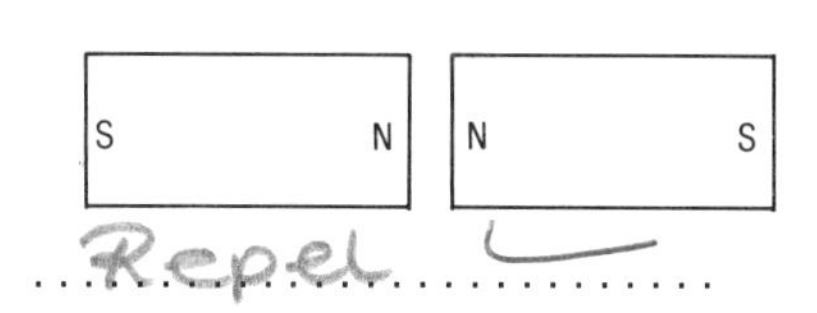

Repel ✓

A2 In a magnetic field, magnetic materials experience a force. ❑

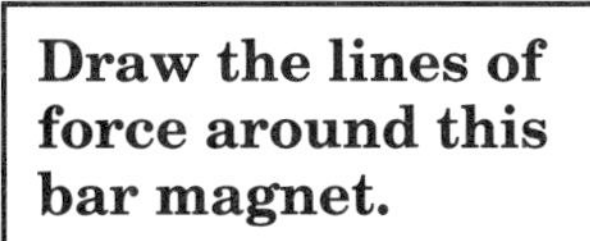

Draw the lines of force around this bar magnet.

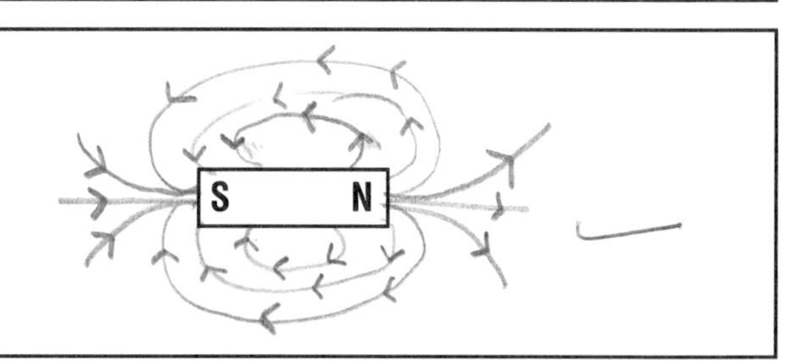

A3 When an electric current flows in a wire, it produces a **magnetic field**. An electromagnet can be produced by coiling wire around an iron rod.

- When the current is on, the iron becomes a strong magnet.
- When the current is switched off, the iron loses its magnetism.
- Electromagnets are used in many everyday appliances such as circuit breakers, transformers and dynamos. ❑

Complete the explanation of how electromagnets are used in these devices. Use the words *spin, magnetised, vibrate, on*.

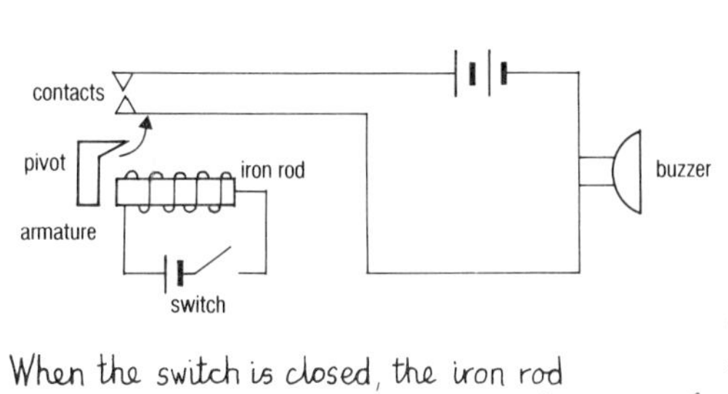

When the switch is closed, the iron rod becomes magnetised ✓
The iron armature is attracted to the iron rod. As it moves it pushes the contacts in the second circuit together.
The buzzer is turned on ✓

Relay

A changing current flows through coil making it move in and out. This makes the cone and the air near it vibrate ✓

Loudspeaker

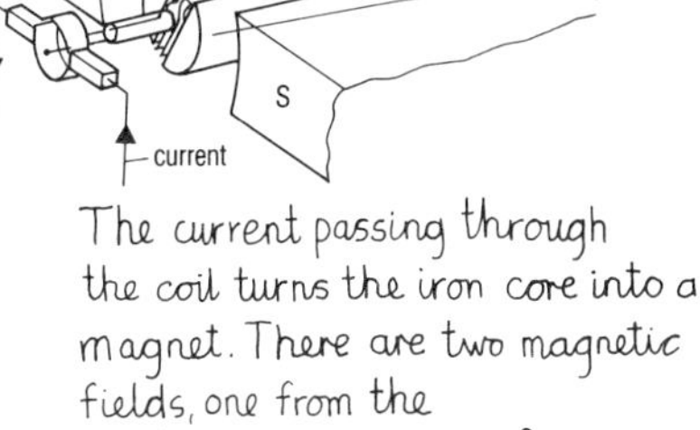

The current passing through the coil turns the iron core into a magnet. There are two magnetic fields, one from the electromagnet and one from the permanent magnet. A force is produced which makes the core spin ✓

Electric motor

B1 Pressure is measured in Pascals and is calculated by dividing the force by the area.

$$\text{pressure (Pa)} = \frac{\text{force (Newtons)}}{\text{area (m}^2\text{)}}$$

❑

Calculate the pressure in the examples.

B2 Speed is calculated from distance travelled and time taken. ☐

What is the average speed of a person who walks 3 km in 45 minutes? Give your answer in metres per second.

B3 Work done is measured in Joules. The formula for work done is:

work = force × distance ☐

Calculate the work done in these examples.

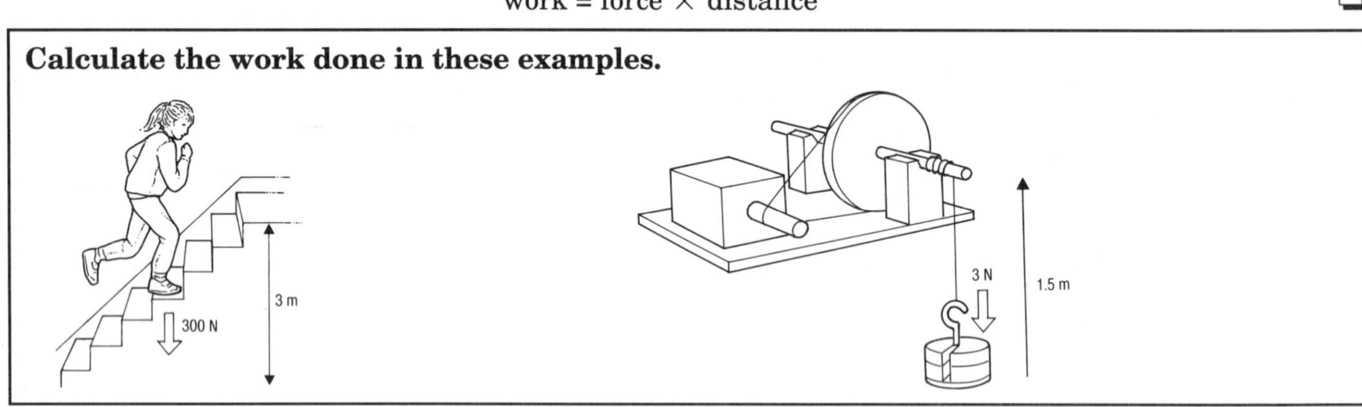

B4 A lever is made up of a bar and a pivot. A force applied at one end of the lever causes a turning effect or **moment**. The turning effect is measured by multiplying the force by the distance *from the pivot*. To balance the turning force, the product of force × distance must be the same on each side of the pivot. For example, the child on the left of the see-saw could be balanced by several different children. ☐

Complete the force × distance calculations.

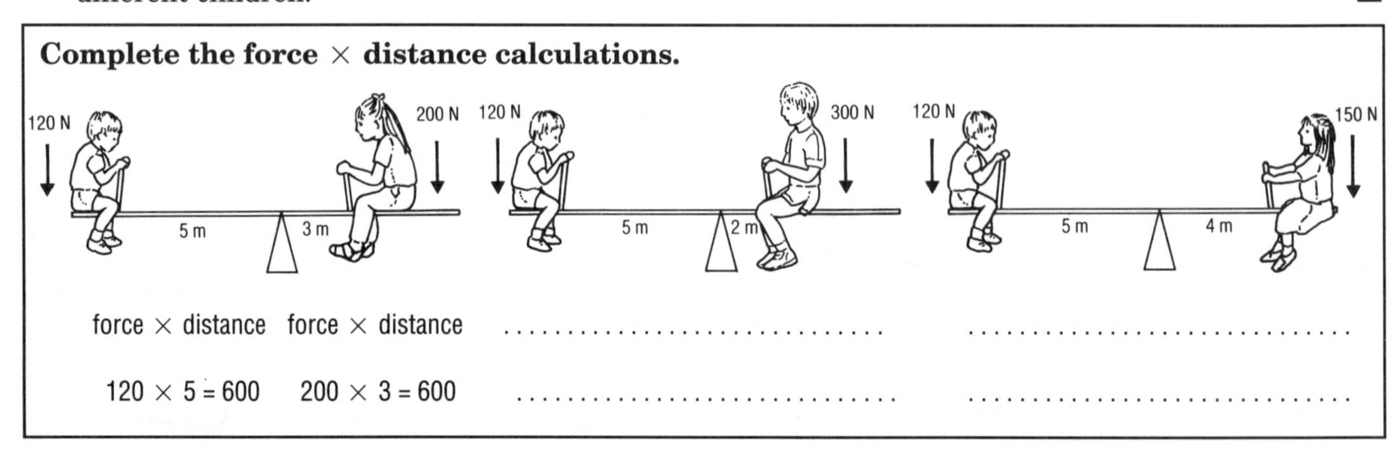

The same idea is used when a crane lifts weights. Complete the calculations. Calculate the distance the load must be from the pivot to keep in balance.

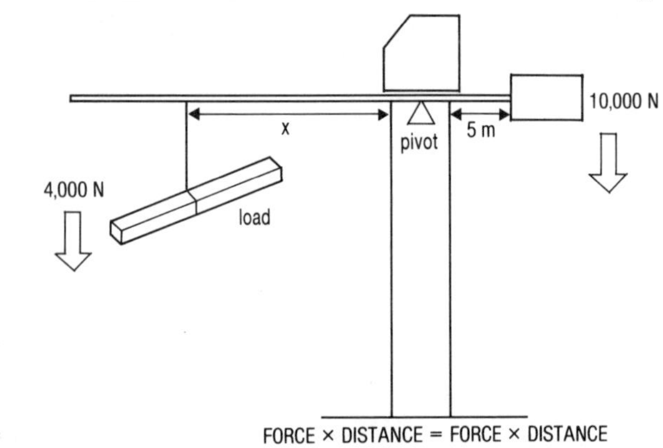

B5 The stability of an object depends on its centre of mass. ❑

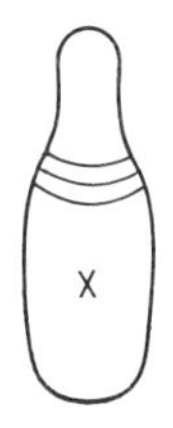

Ten-pin bowling pin

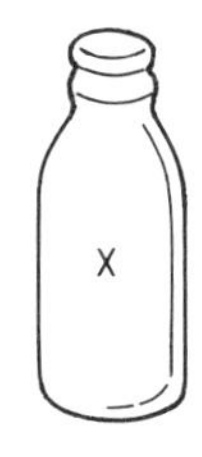

Milk bottle

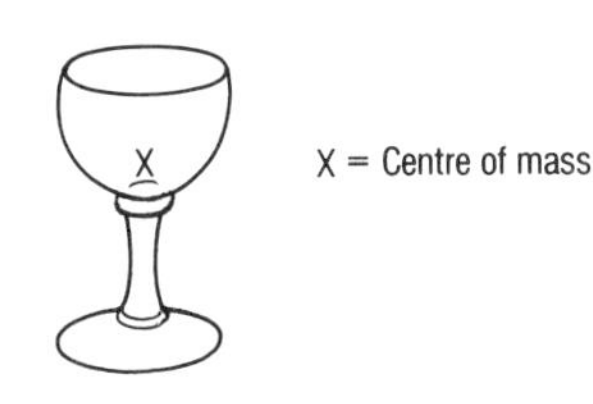

Wine glass

C1 We see objects because light is scattered from them and enters the eye. ❑

Complete the table.

Name of part	Function of part
	lets light in at front of eye
	controls amount of light passing to back of eye
	focuses light
	changes light to nerve impulses
	takes nerve impulses to the brain

C2 White light can be dispersed to give a range of colours. This happens in a rainbow. ❑

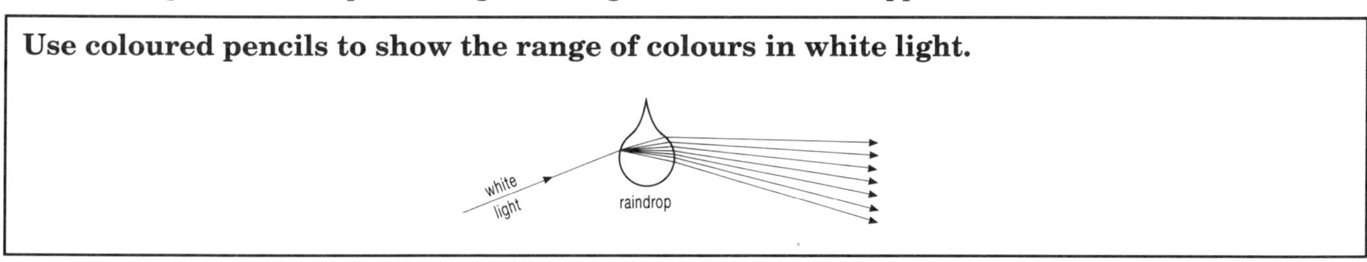

C3 Colour filters absorb all the light except the one you can see. ❑

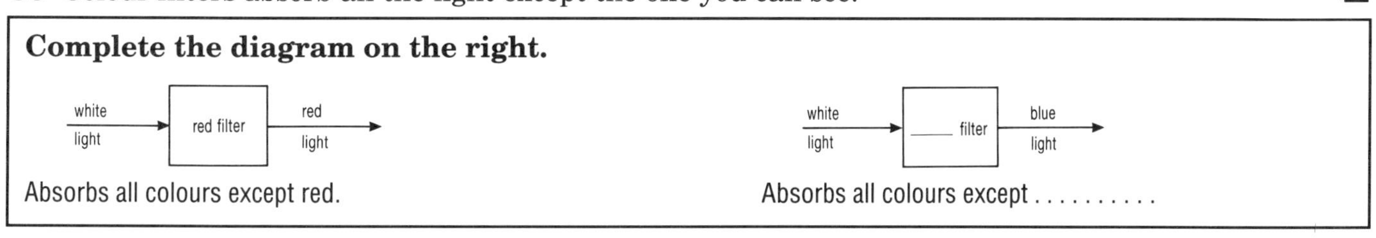

C4 The colour of an object depends on the colour of light which falls on it **and** the colours it reflects. If you put a colour filter between your eye and the object, then its colour will change. ❑

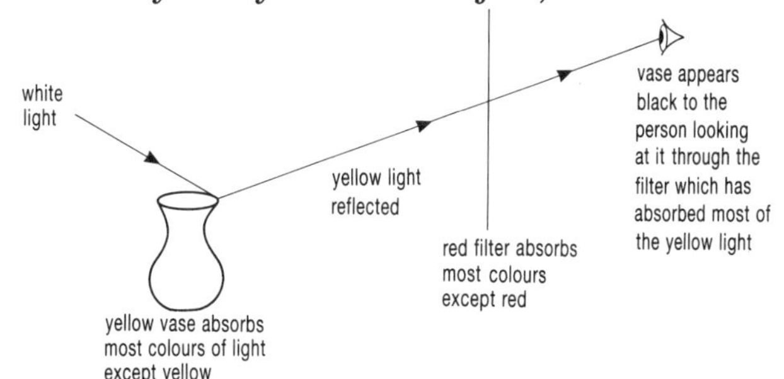

C5 When light strikes a transparent material at right angles, there is no refraction. ❑

C6 Visible light is one kind of electromagnetic radiation. There are many other kinds which make up the electromagnetic spectrum. ☐

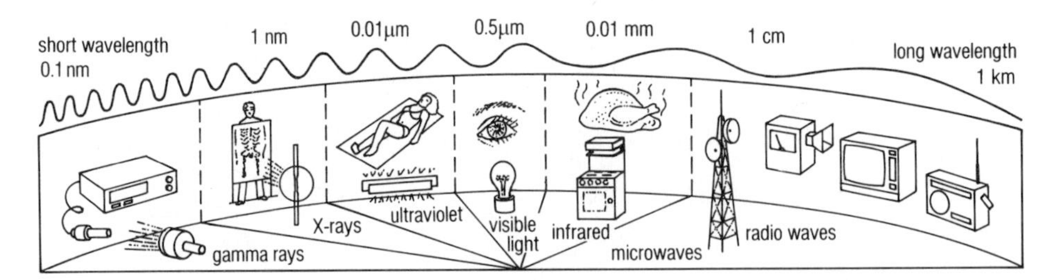

C7 Sound waves can be shown as simple wave diagrams. ☐

Two waves are shown below. Label them to show which is a high note and which is a low note.

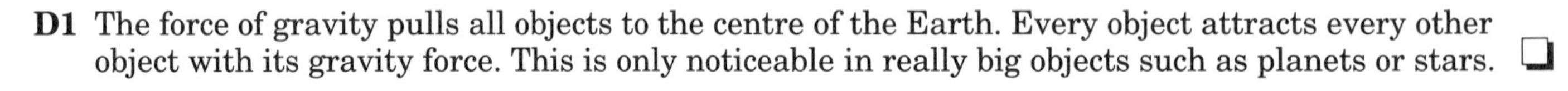

D1 The force of gravity pulls all objects to the centre of the Earth. Every object attracts every other object with its gravity force. This is only noticeable in really big objects such as planets or stars. ☐

D2 The strength of the gravity force depends on the size of the objects and on how far apart the objects are. The larger the objects, the stronger the gravity force. The closer the objects, the stronger the gravity force (also called the **gravity force field**). For example, if a comet entered the solar system, the Sun's gravity field would have little effect at first. As the comet came closer, the Sun's gravity would begin to attract it. The comet would change direction, and might be pulled to the centre of the Sun. ☐

Tick the object in each pair with the stronger gravity force field.

A..... or B..... C..... or D.....

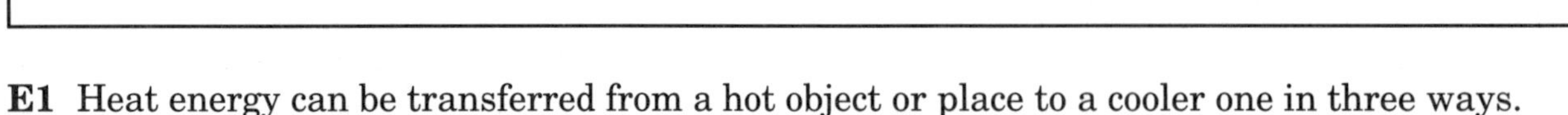

E1 Heat energy can be transferred from a hot object or place to a cooler one in three ways. ☐

- **conduction**
- **convection**
- **radiation**

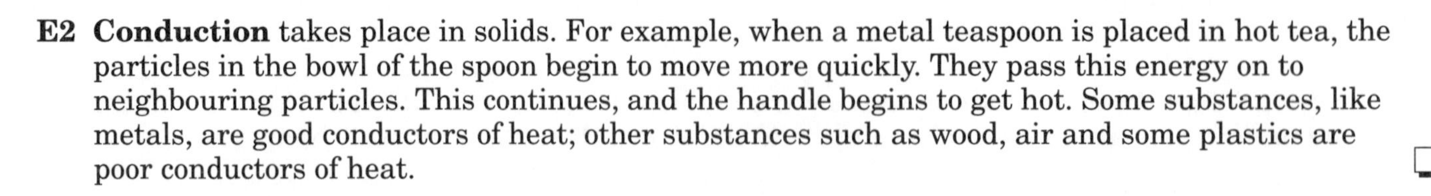

E2 **Conduction** takes place in solids. For example, when a metal teaspoon is placed in hot tea, the particles in the bowl of the spoon begin to move more quickly. They pass this energy on to neighbouring particles. This continues, and the handle begins to get hot. Some substances, like metals, are good conductors of heat; other substances such as wood, air and some plastics are poor conductors of heat. ☐

E3 **Convection** takes place in liquids and gases. For example, in an electric kettle, the particles of warm water move more quickly and become further apart. The warm water becomes less dense. Colder, denser water sinks, and pushes the warm water upwards. This creates convection currents in the water. ☐

E4 **Radiation** takes place in air or space. Hot objects transmit heat energy as waves. For example, heat reaches us from the Sun by radiation.

Decide which method of heat transfer is taking place in these pictures. Write *conduction*, *convection* or *radiation* under each drawing.

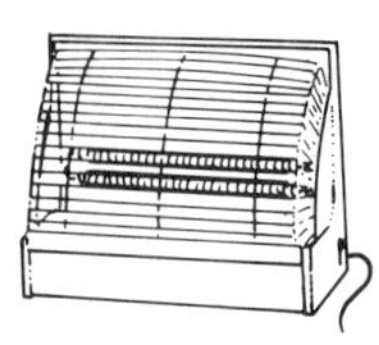 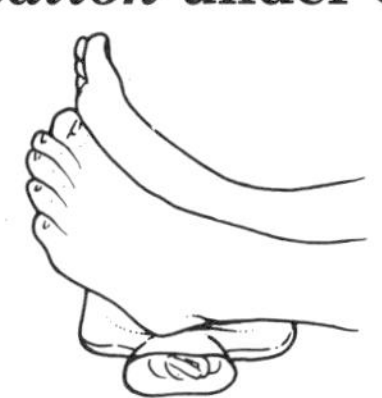  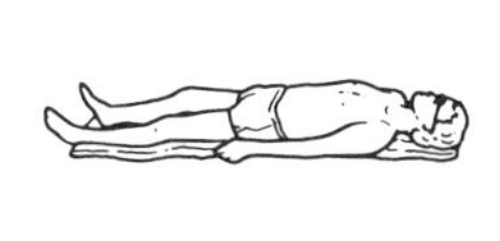 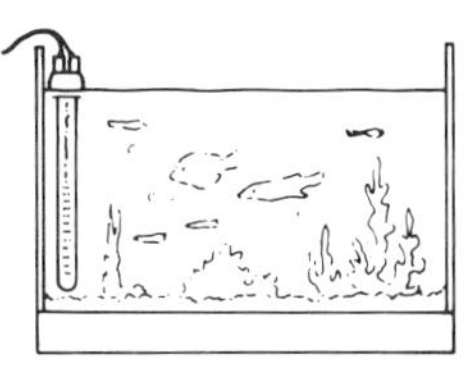

....

E5 A vacuum flask is designed to reduce heat loss.

Complete the diagram by explaining how a vacuum flask works.

Glass container is a poor conductor and prevents heat loss by ....

Shiny surface reflects back any heat and reduces heat loss by ....

There is no air in the vacuum so no heat loss by ....

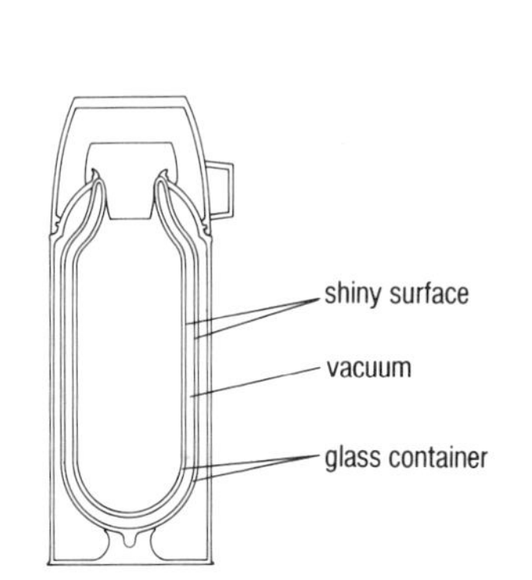

E6 Energy efficiency is measured by comparing how much useful energy is put into a machine, or system, with how much useful energy is put out.

$$\text{energy efficiency} = \frac{\text{energy output}}{\text{energy input}} \times 100 \text{ per cent}$$

The energy efficiency of:
- a pulley is 90%
- a petrol motor is 15%
- human muscle is 40%.

E7 Fuel/oxygen systems provide concentrated sources of energy.

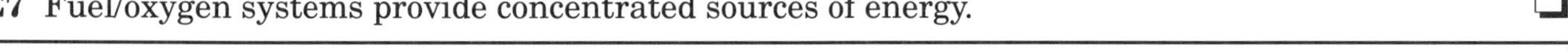

Draw a line to match each fuel/oxygen system with its example.

Food and oxygen	Oil and oxygen	Petrol and oxygen
Central heating system	Car engine	Respiration in living cells

E8 Waste energy is normally released as heat which is spread into the air. This can be trapped and then either used or partly eliminated.
- Waste heat from industrial processes can be transferred to water which is used for heating.
- Friction can be reduced by oiling machinery.

Answers

answers

SC1 encourages investigation, and as a result, there can be several correct responses to a question. The answers to SC1 are a guide only.

SC1 Level 3 (page 8)

A1 an increase in temperature will increase growth.

B1 amount of crowding   the temperature.

SC1 Level 4 (page 9)

A1

presence of draughts	still air is better than draughts
temperature of the surroundings	lower temperature will cause more heat loss.

B1 temperature of the surroundings
type of container used.

C1

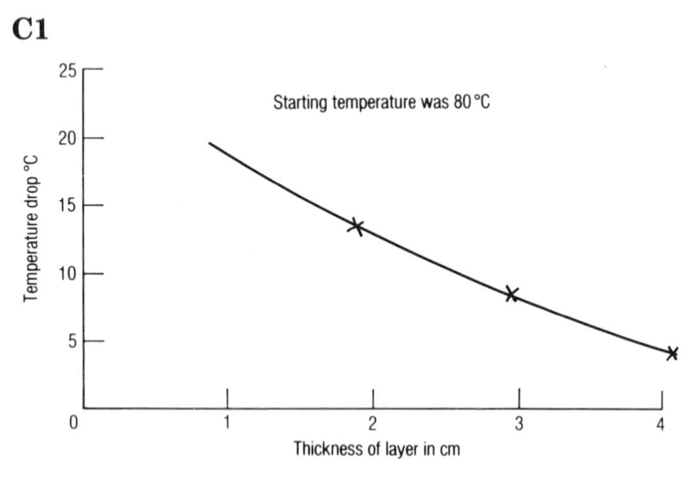

SC1 Level 5 (page 10)

A1

type of yeast used	different yeasts have different rates of reaction
amount of mixing	more mixing will increase rate.

B1 container shape   amount of mixing.

C1

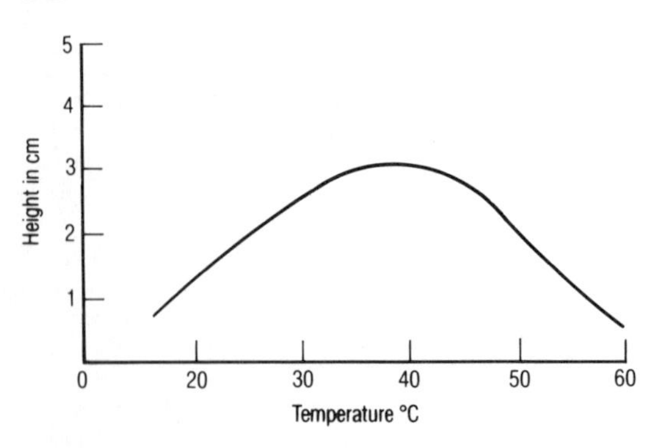

SC1 Level 6 (page 11)

A1 increase rate   increase rate.

B1 particle size   use of a catalyst.

C1

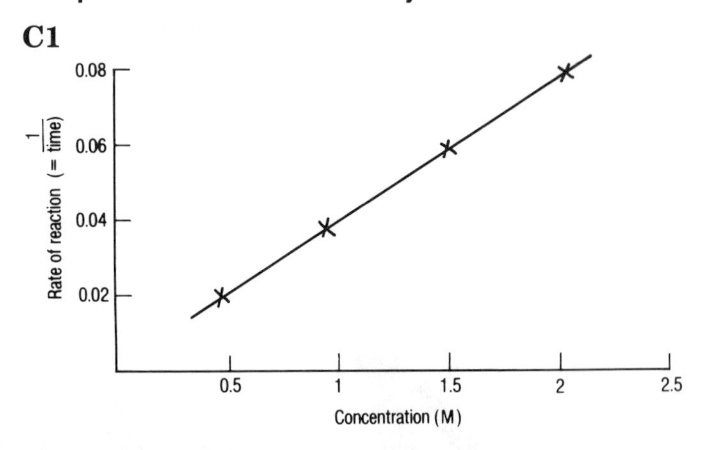

SC1 Level 7 (page 12)

A1 presence of nutrients, for example, nitrogen.

B1 light level   moisture content of soil.

C1 Soil depth has less effect on leaf size than pH and light.

Soil pH and light both affect leaf size.

Leaves are large when the soil pH is 6.

SC2 Level 3 (pages 14 and 15)

A1 a d c b.

B1 reproducing   sensing   feeding   growing   breathing   moving   removing waste.

C1

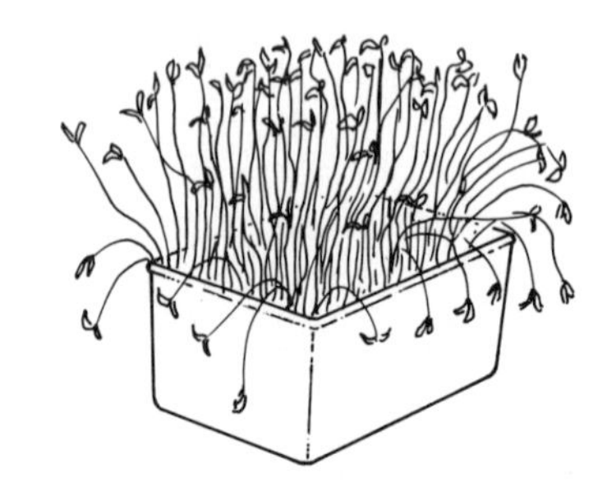

E2

fish	water	fins for swimming
polar bear	freezing temperatures	**thick fur**
camel	dry deserts	**can go without water for a long time**
eg mountain sheep	**mountains**	**good sense of balance**
eg sloth	**trees**	**long, curved claws for hanging from branches**

SC2 Level 4 (pages 16–18)

A2 brain **a**   heart **c**   kidneys **e**   stomach **d**   lungs **b**;   flower **d**   leaf **c**   roots **a**   stamen **e**   ovary **f**   stem **b**.

B2 protein   fat   carbohydrate

B3

protein	**eggs**	**meat**
fat	**fish liver oil**	**cheese**
carbohydrate	**sugar**	**pasta**
minerals	**milk**	**fish**
vitamins	**fresh fruit**	**fresh vegetables**
fibre	**nuts**	**oats**
water	**tea**	**cucumber**

C1 **1 3 4 2.**

D1 **snail earthworm beetle spider.**

D2

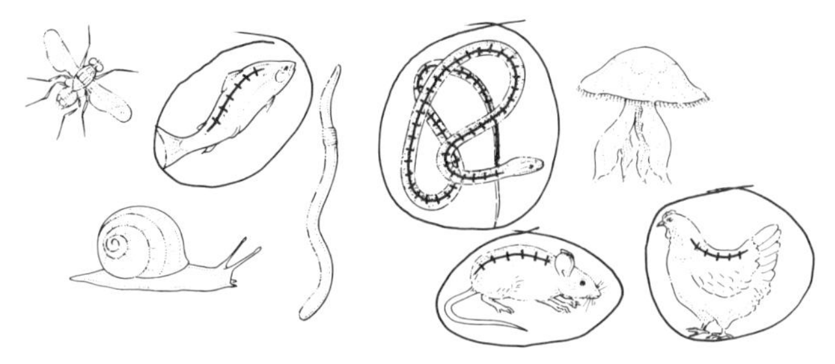

D3 **amphibians birds fish mammals reptiles.**

E1 **water minerals.**

E4 **energy.**

E6

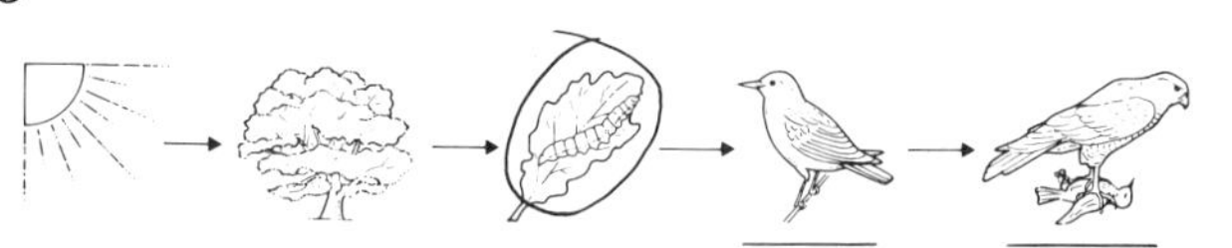

predators – thrush hawk; prey – caterpillar thrush.

SC2 Level 5 (pages 19–21)

B3

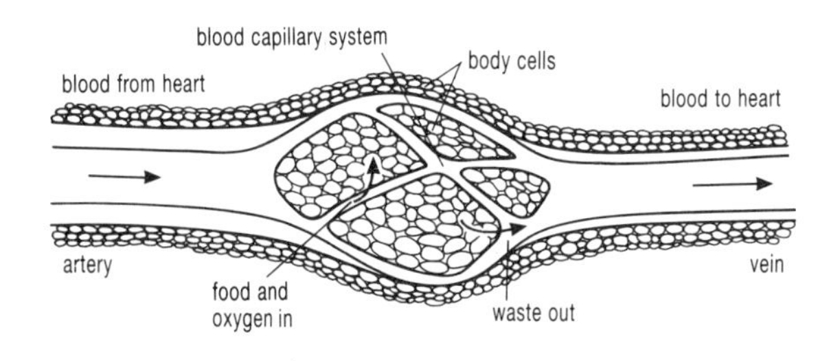

B4 **transport skeletal digestive respiratory nervous reproductive excretory.**

C1

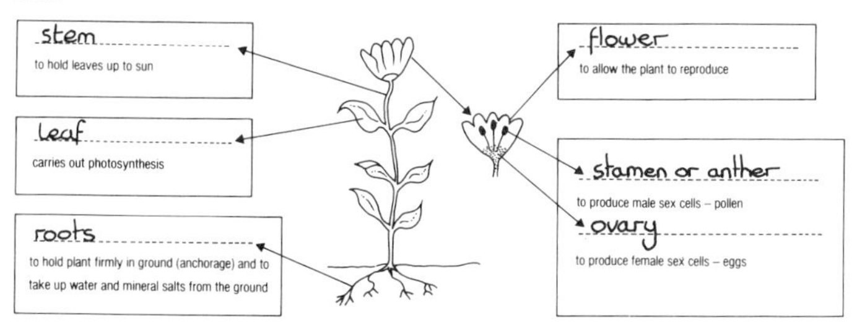

C2

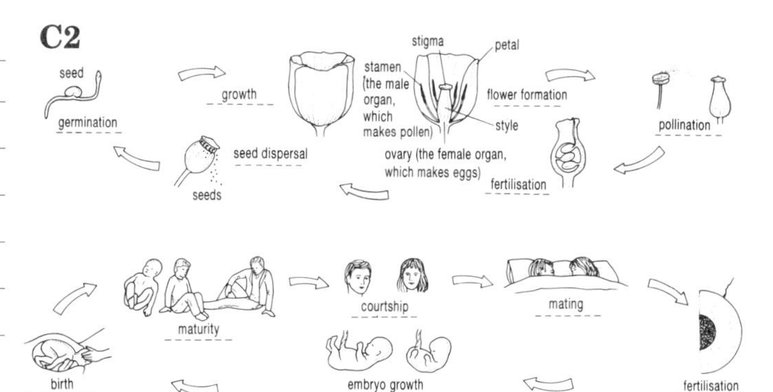

fertilisation and growth.

D1

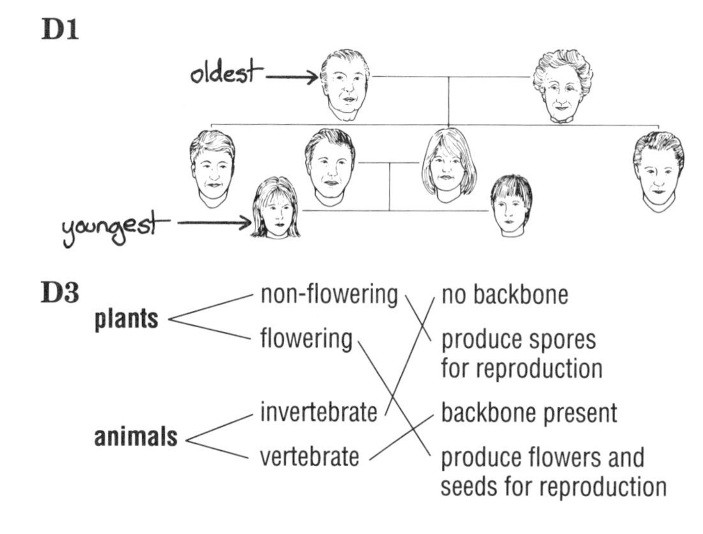

D3

plants – non-flowering – no backbone / produce spores for reproduction
flowering – produce flowers and seeds for reproduction

animals – invertebrate – no backbone
vertebrate – backbone present

D4

fern	**roots and stems**
	leaves divided into leaflets
horsetail	**roots and stems**
	leaves in spirals round stem
mosses and liverwort	**no true roots**
	simple stem with tiny leaves

D5

jellyfish	**hollow body; stinging cells**
flatworm	**flat body; no segments**
arthropod	**hard outer skeleton; jointed legs**
mollusc	**soft body with shell; no segments**
roundworm	**round body; segments**

E1

cactus	desert	very hot, lack of water	**can store water reduced leaves**
polar bear	arctic tundra	very cold	**thick fur**
shark	ocean	oxygen dissolved in water	**gills**
lemur	jungle	**not much light to see**	very big eyes

SC2 Level 6 (pages 22–24)

A1

nucleus
cell membrane
cytoplasm

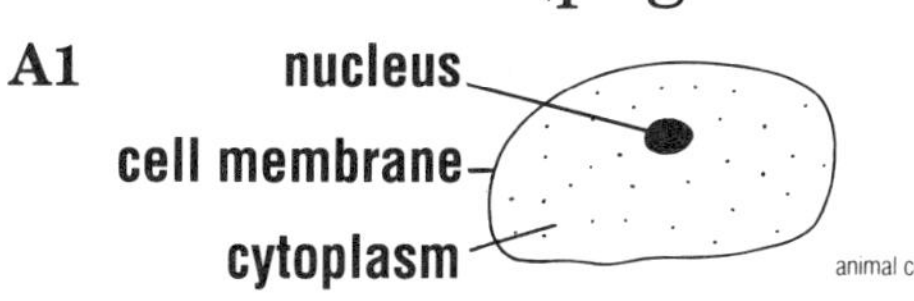

cell wall
chloroplasts
vacuole.

B1 **gas exchange aerobic respiration.**

B3 **mental changes physical changes emotional changes.**

C1 **photosynthesis**
carbon dioxide + **water** —sunlight→ glucose + oxygen

respiration
glucose + oxygen ——→ **carbon dioxide** + **water** + energy

C2 **carbon cycle nitrogen cycle.**

D1 **eye colour blood group right- or left-handed.**

E1 reduce water loss — waterproof covering
reproduce out of water — internal fertilisation and development
breathe air — lungs for gas exchange
support themselves — skeletal system

E4 **increase decrease.**

E5 **Sun → plankton → mussels → starfish;**
Sun → seaweed → sea-snail → crab.

SC2 Level 7 (pages 25–28)

A2 **plant – xylem cell animal – white blood cell.**

B2 **Large surface area for gas exchange.**
Air sacs with thin walls to speed gas exchange.
Air sacs with moist surfaces to speed gas exchange.
Rich blood supply for gas absorption.

B3 **A–3 B–1 C–4 D–2 E–5.**

B4 **deflated inflated.**

B5 large dilute.

B6

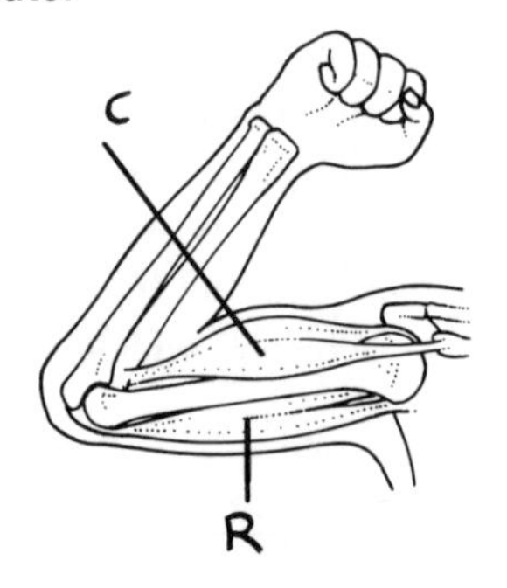

B7 **A–3 B–6 C–1 D–5 E–2 F–4 G–7.**

B8

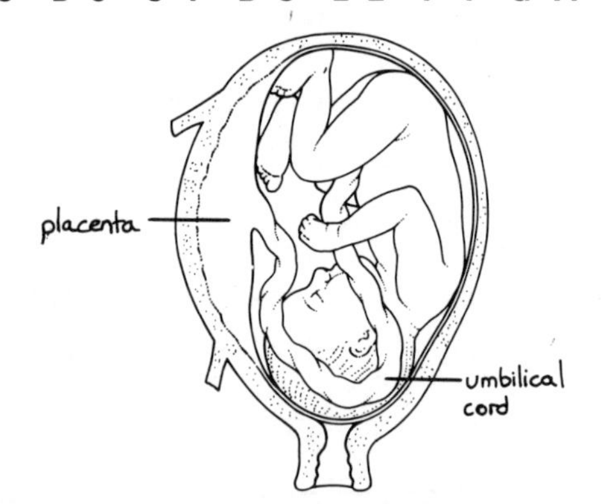

B9

skin	touch	pressure
eye	**sight**	light
ear	hearing	**soundwave**
tongue	**taste**	chemicals
nose	**smell**	chemicals
inner ear	balance	gravity

C1

water + carbon dioxide —light / chlorophyll→ glucose

starch ← glucose: stored

glucose → + oxygen → water + carbon dioxide + energy: aerobic respiration

more complex substances ← glucose: growth and repair

D2 ability to roll your tongue
coat colour in gerbils
height in humans
variation in size of Granny Smith apples
weight in humans

- - - - - **depend on inheritance**

——— **depend on inheritance and environmental factors**

E1

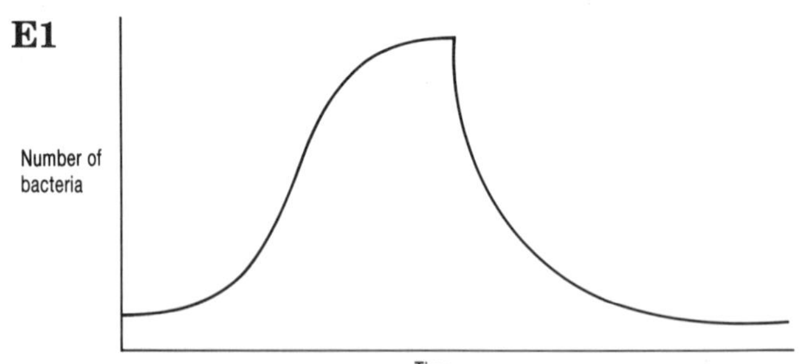

E2 **biomass number.**

SC3 Level 3 (pages 30 and 31)

A2

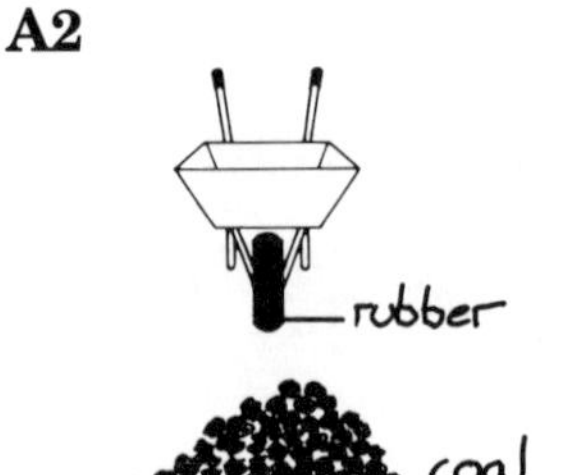

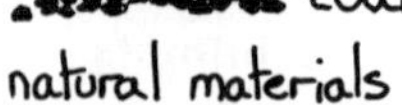

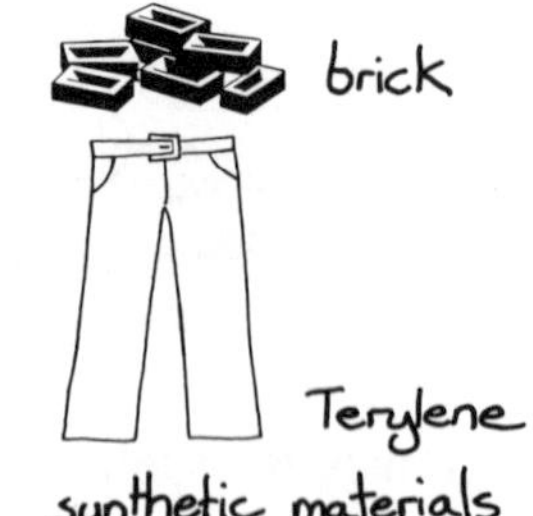

A3

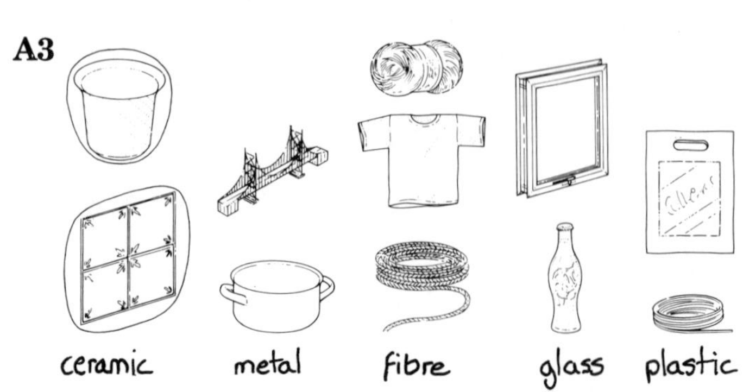

A4

glass	transparent
metal	**electrical conductor**
steel	strong
stone	**durable**
plastic	flexible
rubber	waterproof
lead	dense
balsa wood	**lightweight**

B1 **melting ice lolly   water being spilt in a sugar bowl salt dissolving in soup.**

SC3 Level 4 (pages 32 and 33)

A2

coal	**petrol**	**oxygen**
wood	**water**	**carbon dioxide**
steel	**milk**	**water vapour**
slate	**sulphuric acid**	**hydrogen**
glass	**orange juice**	**air**

A4

A5 sand from a mixture of sand and water — sieving
pebbles from a mixture of pebbles and sand — filtration
water from a mixture of salt and water — evaporation
salt from a mixture of salt and water — condensation

B3 **frying an egg – irreversible**
melting chocolate – reversible.

B4

B6

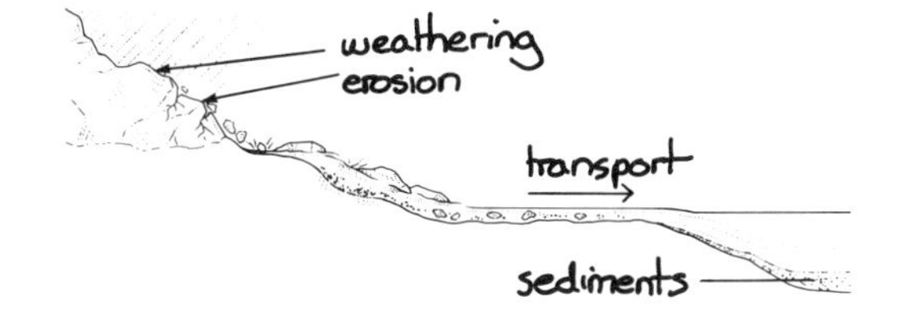

SC3 Level 5 (pages 34–36)

A1

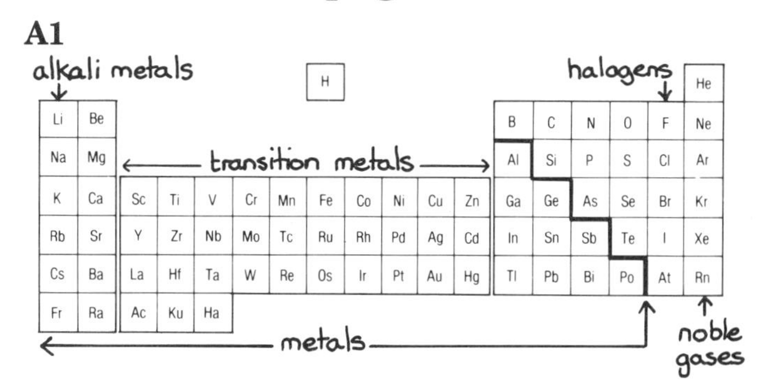

A3 **compounds   mixtures   elements.**

A5 **shiny   conduct heat   drawn out to make wire.**

A6

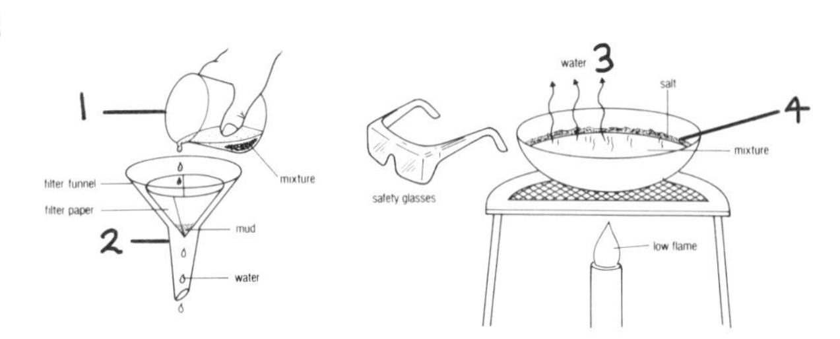

A7 **fuel for heating   fuel for aeroplanes   fuel for cars fuel for ships   lubricants   tar for roads.**

B1

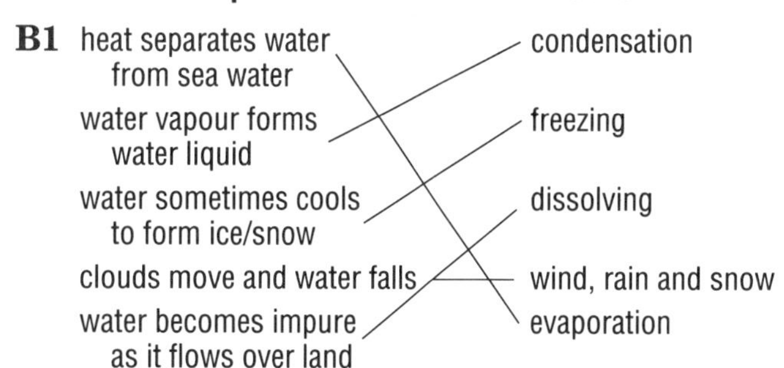

B4

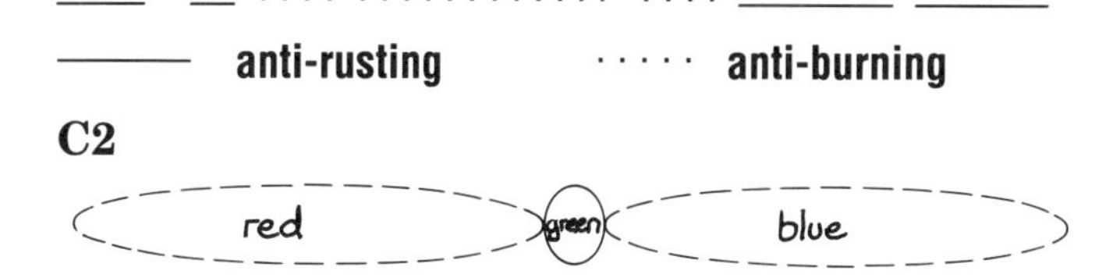

C2

SC3 Level 6 (pages 37–39)

A1

✗	✗
✓	✓
✓	✓
✗	✗
✗	✗
✓	✓

A3

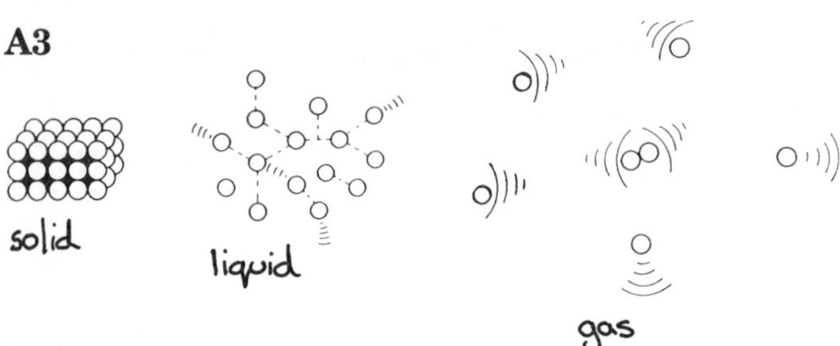

B2 **stays the same.**

B3 copper + oxygen → **copper oxide**
hydrogen + oxygen → **water**
sulphur + **oxygen** → sulphur dioxide
ethanol + oxygen → **carbon dioxide** + **water**.

B5 Na^+ Cl^- O^{2-} Mg^{2+}.

B6 **endothermic exothermic exothermic exothermic.**

B7

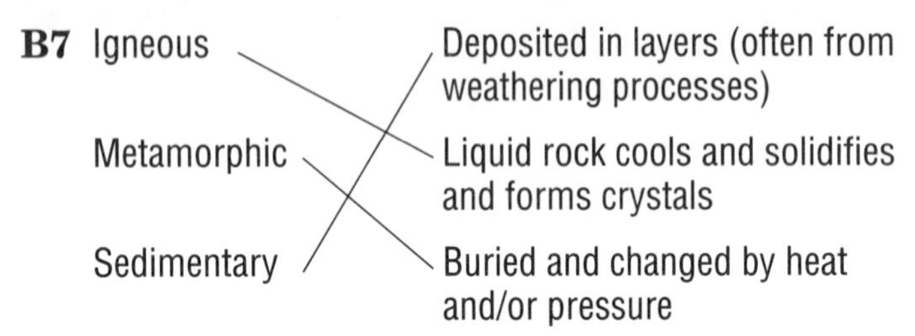

B9 **4 1 3 2.**

C1 **potassium sodium lithium calcium magnesium aluminium zinc iron tin lead hydrogen copper mercury silver gold.**

More energy released **More energy released**

Each word starts with the same letter as the metal name and helps you to remember the order of the reactivity of the metals.

SC3 Level 7 (pages 40–44)

A1

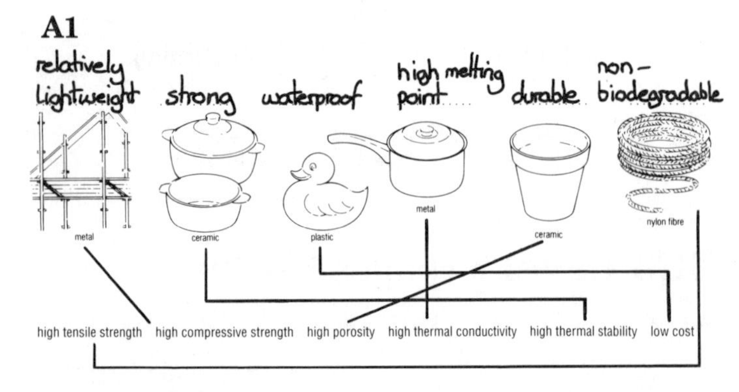

A2 Na **shiny metal, more reactive**

K **shiny metal, more reactive than Na**

Cl **non-metal gas, less reactive**

Br **non-metal liquid, less reactive than Cl**

Ar **non-metal gas, unreactive**

Kr **non-metal gas, unreactive.**

A3

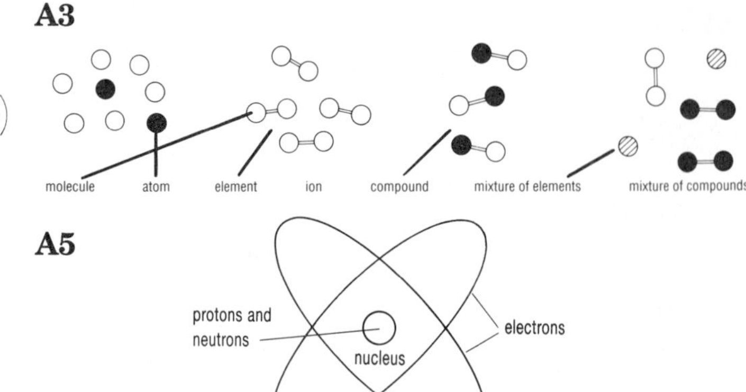

A5

A6 Na – **11** C – **6** Si – **14.**

A7

water	$\mathbf{H_2O}$	**compound**
carbon dioxide	CO_2	**compound**
sodium	Na	**element**
oxygen	O_2	**element**
sodium chloride	**NaCl**	**compound**
iodine	$\mathbf{I_2}$	**element**
magnesium oxide	**MgO**	**compound**

A8 **Particles of orange juice diffuse through water to make orange squash.**

Adding a drop of ink to a tube of water.
Or anything suitable.

A9 **If a substance is heated its particles move more and the space between them increases, leading to a change in state, for example, from a liquid to a gas.**

A10 $$\frac{P_1 \times V_1}{T_1} = \frac{P_2 \times V_2}{T_2}$$

$$\frac{1 \times 3}{290} = \frac{2 \times x}{290}$$

$$2x = \frac{3}{290} \times 290 = 3$$

$$x = 1.5 \text{ litres}$$

B1

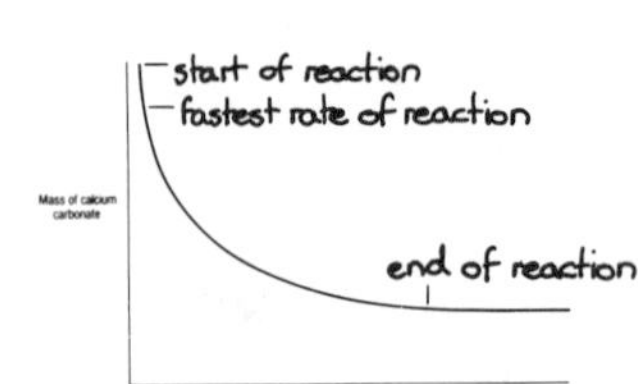

B3 **use of fertiliser** **putting food in an oven**
cutting food **catalytic converter in a car.**

B4 **P1–C; P2–G; P3–A; P4–E**
C1–H; C2–B; C3–D; C4–F.

C1 **7 6 5 2 1 4 3.**

D4

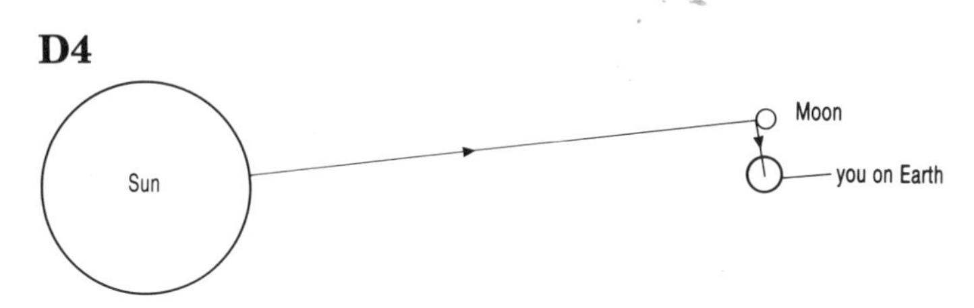

E1 electrical to sound stored to movement
light to electrical to movement.

E2 non-renewable renewable.

SC4 Level 6 (pages 55–58)

A1 low high.

A3 A ✓ B ✓.

A5

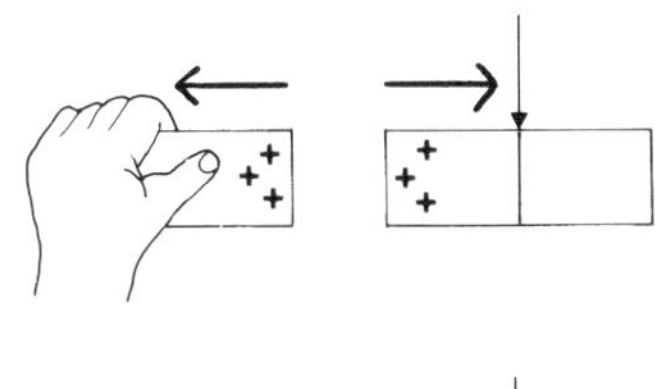

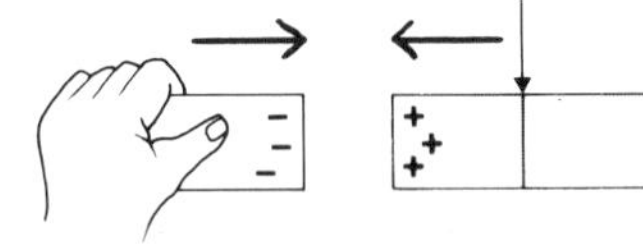

A6 lightning tiny electrical discharges from clothes.

B1

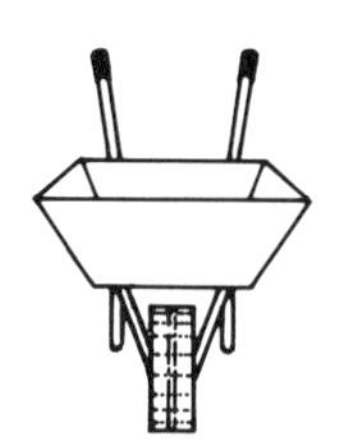

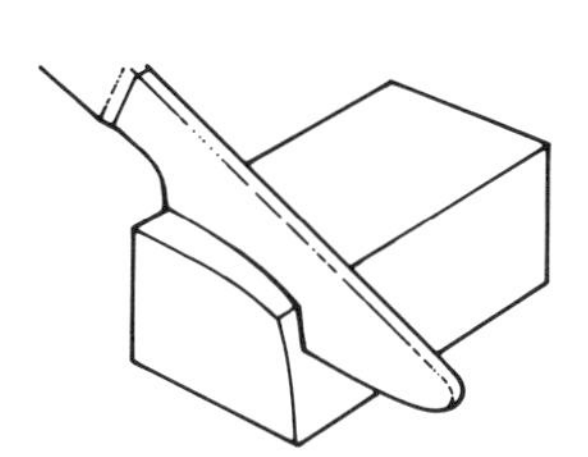

B5 speed = distance/time ✓ $\frac{200}{2.5}$ = 80 km/h

B6 dry deep tread slow rough.

C1 it slows down and is refracted
it slows down and is refracted.

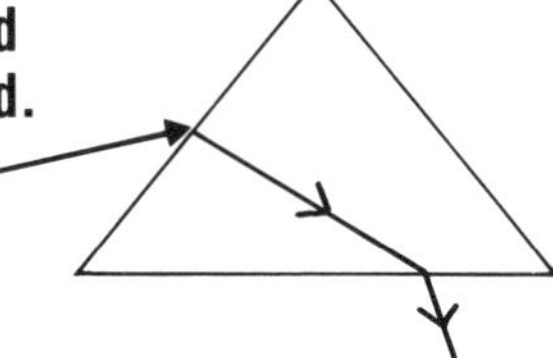

C2 ear drum ear bones cochlea auditory nerve
eustachian tube.

C3 A–loud B–soft C–high D–low.

E1 solar cell guitar electric iron.

E2 heat energy light energy and sound energy.

E4 solar power fossil fuels wind power.

SC4 Level 7 (pages 59–63)

A1 repel attract repel.

A2

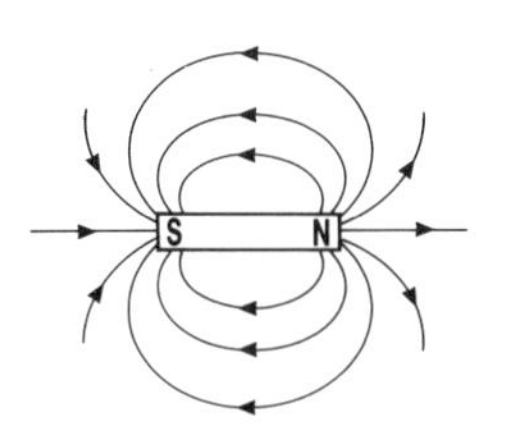

A3 magnetised on vibrate spin.

B1 pressure = $\frac{870}{0.003}$ pressure = $\frac{700}{0.25}$

= 290 000 Pascals = 2800 Pascals

B2 speed = $\frac{3}{0.75}$ speed = $\frac{3000}{45 \times 60}$

= 4 km/h = 1.1 m/s

B3 work = 300 × 3 = 900 Joules work = 3 × 1.5 = 4.5 Joules

B4 120 × 5 = 600 120 × 5 = 600

300 × 2 = 600 150 × 4 = 600

r.h.s. force × distance = 10000 × 5

= 50000

to balance l.h.s.:
force × distance must be equal to 50000

4000 × x = 50000

$x = \frac{50000}{4000}$ = 12.5 m (from the pivot).

C1 cornea iris lens retina optic nerve.

C2

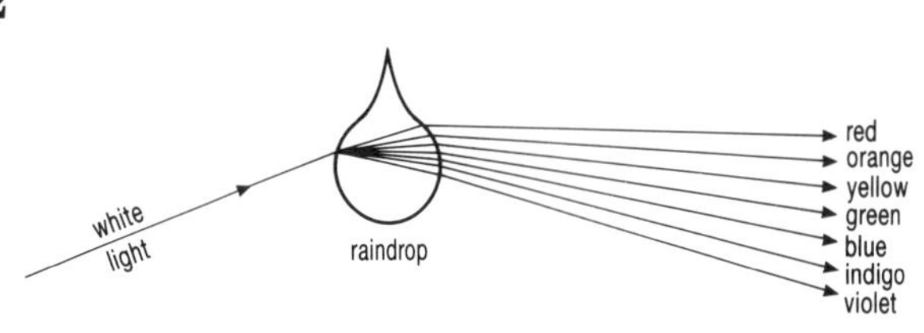

C3 blue.

C7 low high.

D2 A ✓ C ✓.

E4 radiation conduction convection radiation
convection.

E5 conduction radiation convection.

E7

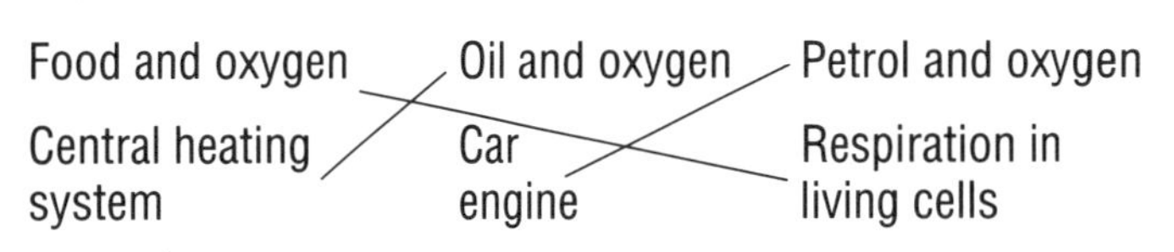

C2

1	2	3	4	5	6	7	8	9	10	11	12	13	14

1 – strong acid, hydrochloric acid; 5 – weak acid, lemon juice; 7 – neutral, water; 11 – weak alkali, bleach; 14 – strong alkali, sodium hydroxide

C3 **sulphuric acid + sodium hydroxide → sodium sulphate + water**

nitric acid + magnesium → magnesium nitrate + hydrogen

C4

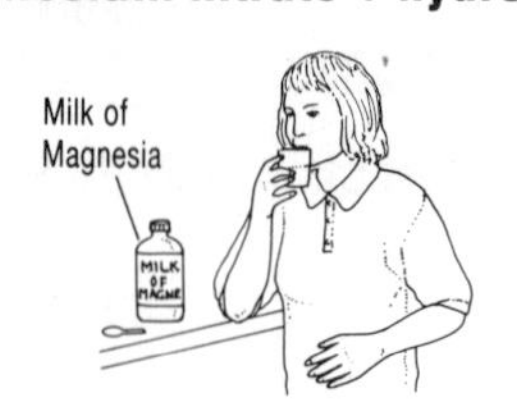

or anything suitable.

SC4 Level 3 (pages 46 and 47)

A1 **insulators conductors.**

A3

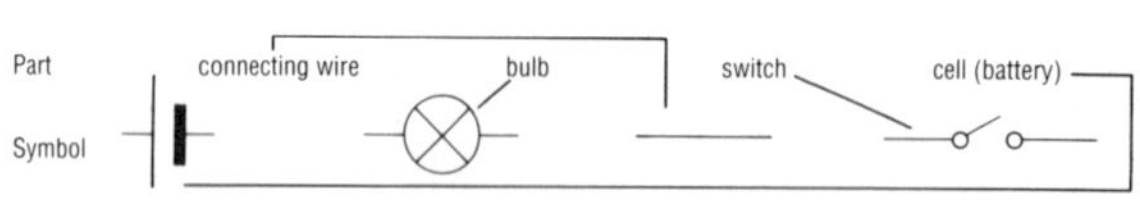

A4 **A D.**

B1 **shape direction speed.**

D1

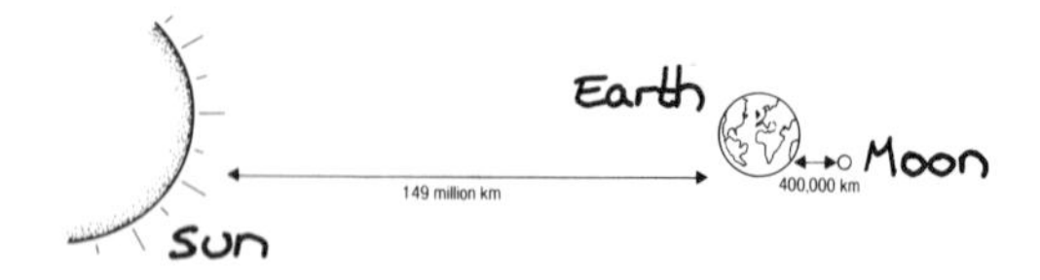

D2

D3 **summer spring winter.**

E2 coal wood gas oil.

SC4 Level 4 (pages 48–50)

A2 **parallel series.**

A3 (a) **A** (b) **A and B.**

B3

pushing force stops as soon as cannon ball leaves the cannon

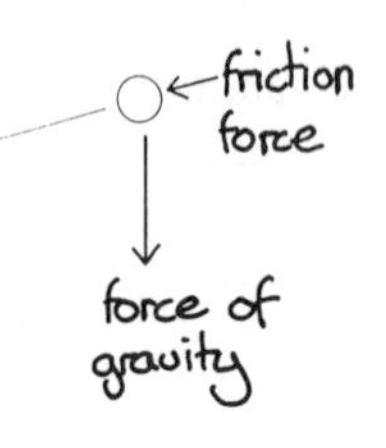

C1 **before after.**

D3 **summer, long days winter, short days.**

D4 **long short.**

E2 **movement; electrical to chemical; movement; electrical to movement.**

SC4 Level 5 (pages 51–54)

A3

relay	turns a current on or off when pushed or pressed
resistor	reduces the current by a fixed amount
switch	a switch operated by a current
variable resistor	controls the range of current that can flow

B1

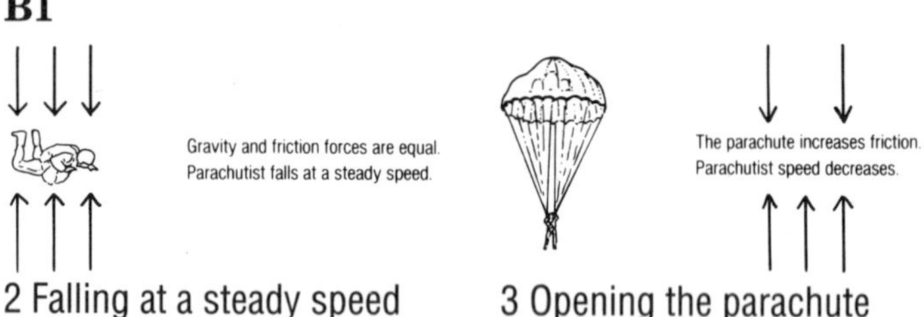

C1

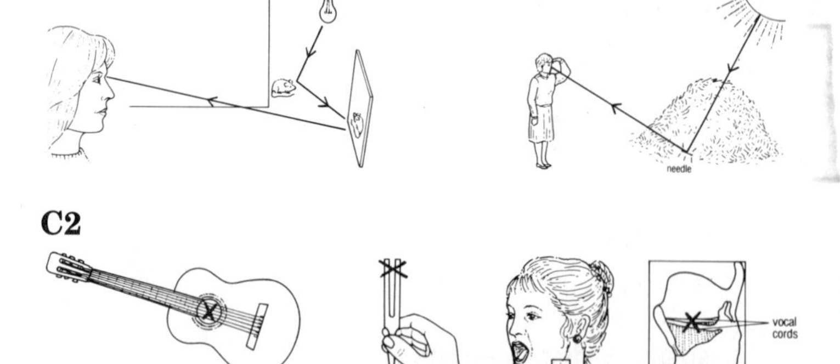

C2

C3 **Ruler (a) pulling it down further**
(b) making the vibrating part shorter by pulling it further across the table.

Milk bottle (a) blowing harder
(b) filling it up more with water.

D1

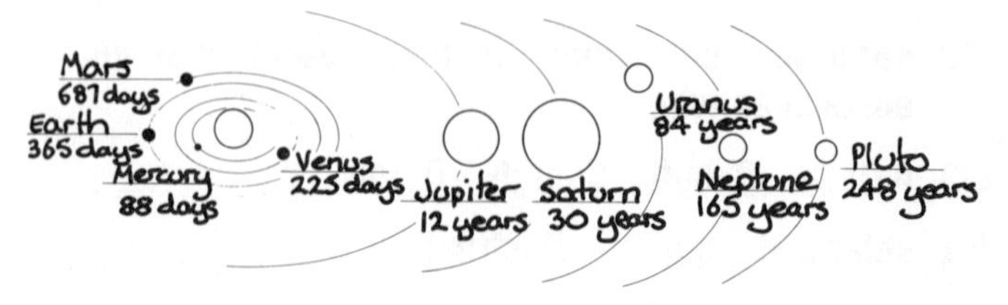

D3 **summer/day winter/night summer/night winter/day.**